THE FERRARI UNDER THE BED

A Memoir and History of 0046M

DARRELL WESTFAUL

DALTON WATSON FINE BOOKS

THE FERRARI UNDER THE BED

A Memoir and History of 0046M

DARRELL WESTFAUL

ISBN: 978-1-956309-23-2

Printed by Interpress Ltd, Hungary

for the publisher

Dalton Watson Fine Books
Glyn and Jean Morris
Deerfield, IL 60015 USA
www.daltonwatson.com

Cover photograph (modern color with author) by Chip Cooper, background photograph courtesy of Monica Valeriani

The author in the driver's seat of 0046M. *Chip Cooper*

For Ary

CONTENTS

2016 Pebble Beach Auction: 1950 Ferrari 166 MM Berlinetta. Copyright and courtesy of Gooding & Company

PREFACE

By sheer luck and a little determination, I had the honor of owning a rare, historic Ferrari and kept it for 40 years. This book is the result of a half-century of dedication and off-and-on research to learn what it was, where it came from, and who had owned it before me.

Time has scattered some of 0046M's early history, but I have managed to piece together enough of the puzzle to tell its story. To be sure, there are pieces missing. Those may be found some day; meanwhile, in places I have guessed and speculated about what might have been in order to understand why things happened as they did.

This book is more than a collection of records and dates and about more than a car. It is also about the people through whose lives and hands 0046M passed before it reached me. Almost invariably, it was a passion of youth, with most of us buying it in our twenties. It appealed to us in a visceral way – a fast car that looked good – and for those of us in its post-racing life, it touched us to the degree that we were willing to accept its impracticality. Indeed, that was an attraction.

I was the last owner for whom 0046M had served the purpose of transportation, and all those before me are now gone. After me, it entered the sheltered life of a collectible and as such it will still outlast each new owner in turn, each of those having his – or her – story joined with the car as well. Barring a catastrophe, it has the immortality of art, transformed from utility to something that will speak of us through time.

Darrell Westfaul
July 2025

Chip Cooper

ACKNOWLEDGMENTS

There are many people whom I wish to thank for their assistance in the research and writing of this book. Many of them are mentioned in the following chapters, but I feel it appropriate to single out those who have been most helpful. It goes without saying that previous owners and their wives I interviewed were invaluable, as well as Alexis Callier and Louel Gibbons who made editorial contributions. Others were as follows:

Alberto Mario Pintus
Alessandro Silva
Antoine Prunet
Bill Green
Boudewijn Berkhoff
Cyril Teste
Daniel M. DiLandro, Archivist, SUNY Buffalo State Library
Dick Merritt
Foto Locchi
Francesco Reggiani
Fred Leydorf
Gerald Roush
Giorgio Nada
Giovanni Zimotti
James S. Leopold
Jess Pourret
John Apen
Laila Quintano
Lorenzo Boscarelli
Luigi Orsini
Marcel Massini
Mary Virginia Reese
Michele Marchianò
Michael Lynch
Monica Valeriani
Patrick and Tazio Ottis
Stan Nowak
The Klemantaski Collection
The Revs Institute
Walter Bäumer
Zagato Historical Archives

I also want to express my gratitude to Andreas Pohl for his generosity in supporting the publication of this book, to Rainer Wolf for his facilitation of the project, to Glyn Morris of Dalton Watson Fine Books for his patience, to Jodi Ellis for her graphic skills, and to Andrea Zagato for the courtesy of his Foreword.

FOREWORD

Zagato never collaborated directly with Ferrari Automobili, but the strong relationship between their founders, Ugo Zagato and Enzo Ferrari, dates back to 1921, when they first met in Milan. Originally, Enzo was an Alfa Romeo test driver, and after seeing the results of the light weight Zagato construction, he chose my grandfather's coachworks for racing. Eventually, Enzo became an Alfa Romeo dealer, which led him to found Scuderia Ferrari at Modena in 1929, and he chose Carrozzeria Zagato as an exclusive partner.[1]

In 1947, Enzo founded Ferrari Automobili and the first (patented) design he received was from my grandfather for an 1100 Ferrari Panoramica and an 1100 Ferrari Spider. That illustrates an extraordinary story, a tradition of tribute that has been renewed by my family every decade for over 100 years. Without a direct collaboration with Ferrari, at least 45 different models of Scuderia Ferrari and Ferrari can be counted (23 for Scuderia Ferrari and 22 as tributes to Enzo Ferrari). These numbers include the first street legal Enzo Ferrari car in history (1934) and the two mentioned 1100 drawings – never built, but used as inspiration for the 166 Panoramica of Antonio Stagnoli and the 166 MM of Luigi Bosisio.

Andrea Michele Zagato

The 1953 166 MM Zagato is therefore a tribute to Enzo made by my family. The enormous research done in order to create this book is a testament to a story that continues to inspire and deserves to be told.

Great stories never end!

Andrea Michele Zagato
President, ZED Milano s.r.l.

This rare 1920s photograph is believed to be the only known image capturing Ugo Zagato and Enzo Ferrari together. Top left: Ugo Zagato, top middle: Antonio Ascari, on his left in the hat is Vittorio Jano and Enzo Ferrari is seated far right. *Archivio Zagato ©*

[1] The 1929 Scuderia Ferrari was organized for the private racing of Alfa Romeo cars. In 1937, Alfa bought controlling interest and then liquidated the team at the end of the year, keeping Enzo to run the factory team, then fired him in 1939.

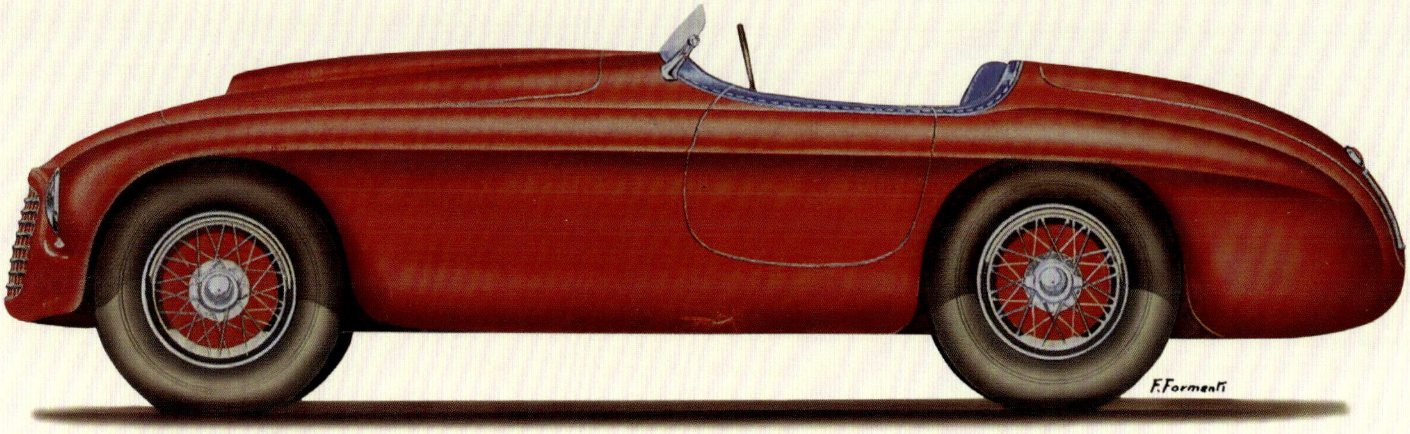

166 Mille Miglia barchetta design study by Carrozzeria Touring.

Artist's rendition of 0046M Zagato berlinetta. *Archivio Zagato ©*

Chapter 1

THE FERRARI UNDER THE BED

0046M

It was 1965, on a bright spring day at the University of Alabama. I was 20 years old, at the close of my sophomore year, and I was walking down the front steps of the student union building, which was located on the southwest corner of University Boulevard and Colonial Drive. Even now, I can see it as clearly as I did back then – a little red car like none I had ever seen before was stopped facing me at the light, diagonally across on University Boulevard. It sat looking aggressive with a long hood followed by a tiny cabin behind a split windshield. The light changed and the car went past, westbound down the street. It was only a quick glimpse, but I was sure I had seen a Ferrari.

I was no stranger to Ferraris. In my high school years in Mobile, Alabama, I had hung out at Joe Smith's garage where he repaired foreign cars and maintained race cars for Ford dealer Joe Treadwell and others. As a pro bono helper, I had put my hands on a 250F Maserati, a Ferrari Mondial, a Lotus Formula Junior, and many other exotic machines that passed through. With that entrée, I had become a Sports Car Club of America (SCCA) corner worker at races when I was 18 and freely roamed the pits at two US Road Racing Championship races in Pensacola and even at the 12 Hours of Sebring in 1963, to where I had bummed a ride in a 3500 GT Maserati. I was a teenage kid among Ferrari GTOs and Cobras while my peers rolled in '57 Chevys.

When I went on to college at Tuscaloosa in the fall of 1963, I left Mobile and motorsports behind, but my early exposure to sports cars spoiled me for ordinary transportation, which was why I had yet to own a car. There were some I had looked at and could afford but did not want, like the Jowett Jupiter for $300, and there were some that I had looked at and wanted but could not afford, like the '54 Corvette for $800.[2] It was more important to scrape up the money to go to college, though, so in the meantime I dreamed of having something like the little red car I had seen.

When I got back to my dorm, I told the guys on my floor what I had seen. In a stroke of luck, one of them had seen the car I described parked at a house maybe five miles (8 km) away, on a side street off Highway 11, the pre-Interstate road to Birmingham. I had to satisfy my curiosity, so the next Saturday my best friend Conrad Harwell and I borrowed a pair of bicycles and set out to find it. It seemed like we pedaled forever, but finally we came to the side road that had been described, went up a hill to the first house on the left – and there it was, sitting in the front yard.

Up close, it was tiny, the neatest little thing, with a huge quick-release gas cap in the rear window, and it was right-hand drive! I fell in love. Charles Deerman, the owner, about 30, came to the door after I knocked and answered my questions about the car. It was indeed a Ferrari, but it now had a Chevy engine. He had bought it in Birmingham and his belief of its origins was that it was a 1954 and one of three cars built for the Mexican Road Race (Carrera Panamericana), a fact easily disproven later, but it illustrates the kind of mistaken lore old Ferraris accrued in those days. Then I asked if he wanted to sell it. He said no, because he had children and thought the car would be something fun for them to have when they were older. I had no money anyway, so we thanked him for letting us look and began our long bike ride back to the university.

As we rode, I said to Conrad, "I want that car."

Later on, I was telling the story of finding a Ferrari in Tuscaloosa to the guys back at the dorm again when there was another stroke of luck. Richard Haskell spoke up, saying that he knew a man back in Birmingham who had a Ferrari engine under his bed. There was scoffing and disbelief, of course, but Richard swore that it was true. If it was, I thought there might be a chance to get that engine to replace the Chevy motor in the Ferrari I planned to buy someday. So, I arranged with Richard for a trip to Birmingham to see the Ferrari engine.

I was skeptical about what I would find, but Richard and I went to the home of Bob Rabun who led us into his bedroom. Sure enough, Bob proceeded to pull, from under his bed, box after box filled with what were undoubtedly the parts

[2] Respectively, the values would be $3,100 and $8,250 in 2024 dollars, without considering appreciation. These and all subsequent inflation calculations are from: www.usinflationcalculator.com, accessed January 15, 2025.

[1] Bob Rabun, who had parts of the Ferrari engine under his bed, and his Cobra. *Author*

of a Ferrari engine: the polished aluminum camshaft covers with the name embossed on them, twelve little pistons and rods, heads, three dual-throat Webers, and on and on. In a closet was the block and the crankshaft was standing upright bolted to the flywheel. It was amazing.

Bob explained that he and a partner had bought the engine and transmission with the intention of building a race car. They were not told, however, that the engine had dropped a valve and seized. When they discovered that, they set about repairing the damage locally, but ultimately they abandoned the project, splitting the engine and transmission between them, with Bob getting the engine. And there it had sat in pieces, under his bed.

Bob was agreeable to selling the engine to me, but as a college student scraping by, there was still the problem of my lack of money. Fortunately, I was eventually able to make a deal on August 2nd: Bob would keep the engine in his possession and I would pay him $50 down and $30 a month until he had received one half of the agreed price, whereupon I could pick it up and continue the payments until he had received the total price of $375.[3] The bill of sale is in Appendix 8. We each kept our bargain, and during the 1965 Christmas break, Conrad and I drove up to Birmingham from Mobile in his 1956 Chevy with the back seat removed to hold it all, and brought back my prize. By then Bob had his own prize, a new Cobra [1].

After we had loaded up the engine, I asked Bob about the transmission, hoping that I could get it too. He explained that his partner had already put it in another car but he could take me to see it, so we drove over. The partner was not there but the car was – it was a Lotus 11. He had installed a flat-six Corvair rear engine in the *front* of the Lotus and mounted

[3] In 2024 dollars: $500 down, $300 per month until a total of $3,760.

the Ferrari transmission behind it. To do so, he had cut off the bell housing. That damage and the fact that it was in use caused me to lose interest until years later when I would try to find the car in order to recover the transmission. My thinking then was that I could find some other transmission that would work, perhaps from an Alfa Romeo, so I was not too concerned.

At the time, I never questioned Bob about the source of the engine, but since there was an engine without a car and a car without an engine, I suspected that the one I had bought might indeed be the one missing from the Ferrari in Tuscaloosa. Serial numbers and Ferrari production were still largely a mysterious subject then, into which I had no insight. Anyway, I was overjoyed to hold in my hands the hidden secrets of a Ferrari – beautifully cast, polished or machined parts, all numbered for their specific cylinder. Even the oil dipstick was machined from a solid billet of steel. Squirreling them away for the future, I packed the parts of the engine in three boxes, one that I made of plywood to fit the block, and two that were antiquities in themselves – World War II fragmentation bomb crates that were sturdy and had rope handles on each end **[2-3]**. I lined all the boxes with roofing tar paper to seal out moisture and I coated the parts with oil and grease. That preserved them well for an unexpected forty years.

Of course, the car was the real prize I sought, so I saved money from summer jobs and added that to a considerable contribution from my parents until I had put together $1,500 by the summer of 1966.[4] That I hoped would be enough, and so I drove my father's 1957 Plymouth to Tuscaloosa on June 28 to see Deerman. I had had no contact with him in over a year, such was my blind – and naïve – faith that the Ferrari was destined to be mine, so my heart sank when I saw that it was missing when I drove up. Devastated, I went to his door anyway. He was home, and he remembered me, but best of all he still had the Ferrari. It was in Columbus, Mississippi, for painting and I feared that meant he intended to keep it. But, I asked if he would consider selling now and he said he

[2] Engine block 0046M stored in my home-made crate. *Author*

[3] Miscellaneous 0046M engine parts stored in World War II bomb crates. *Author*

would; the question was for how much. I had no experience with negotiation; all I knew was that I had come to buy the car, so I offered all I had, the entire $1,500, and he took it. Four decades later, when I saw him for the second and last time, Deerman showed me his own bill of sale for the Ferrari, and it turned out that he had paid $1,500 for the car himself four years before. Since there was no loss for him, he had probably thought it was a good deal.

4 In 2024 dollars: $14,600.

I spent the night with my friend Conrad who had married and moved into an apartment, and the following day Deerman and I set out in his car for Columbus to see the Ferrari. At the painter's, I was in for a shock. Not understanding the Italian manner of constructing hand-built aluminum bodies, someone had sandblasted it, which removed much of the filler that had been liberally applied to smooth the rough hammered metal, leaving the car motley-colored in dull bare aluminum with splotches of pitted gray and pink filler. It was ugly. Nevertheless, I was committed – if a little dubious – even after seeing the car's condition. It was raining that day, and the car's windshield wipers were not working, so we returned to Tuscaloosa without it.

The next day, June 30, we returned and drove the car back. That is, Deerman drove with his wife following in his car. As much as I wanted it, I was timid about trying the right-hand drive, so I made the first trip in my Ferrari as a passenger. Along the way, it began to overheat and coolant was spraying from a couple of pinholes in the top of the radiator, running through the louvers in the front of the hood, and sprinkling the windshield. We stopped, let the engine cool down, and took it easy the rest of the way with no trouble. Even so, I had a preview of what I was getting myself into. It was summer in Alabama, a sunny, humid day, and there was a hot engine pressed against a thin firewall with sparse insulation, so it was broiling hot in the car, especially since the Plexiglass door windows were fixed in place with only vent windows in the doors and rear quarter windows that swung out a few inches. It was also cramped, loud and rode hard, but I was 21 years old, buying my first car – and it was a Ferrari!

We parked the Ferrari back in Deerman's front yard, and since I had driven my dad's car to Tuscaloosa, I returned to Mobile in it. A few days later, on the Fourth of July, I went back to Tuscaloosa with Conrad and his wife who had come home for a visit. The next day, Conrad drove me to Deerman's house and I exchanged money for the Ferrari's bill of sale with Conrad as witness (Appendix 9). Really just as an afterthought, it occurred to me to ask before I left if he had any other parts to the car. He told me to wait and disappeared around the back of his house. He returned carrying a dirty folded sheet of aluminum, which was the belly pan, one muddy axle tube that turned out to be for the left side, and the original differential, now packed full of red clay from being tossed into the gully back there. I already had the long aluminum trim pieces stuffed inside the car, but we wedged the other parts into the car as well, the folded belly pan going over the spare tire on the shelf behind the seats. At the time, neither of us had any idea how rare those parts were, and it was simply luck that they were saved. Many years later, I wondered if there might have been more down in that gully, but I never went back to look and a shopping center sits there now.

With some trepidation, it was time for me to leave and drive the car for the first time. I folded my six-foot frame into the driver's seat in a car designed for short people, but at least I was thin. It felt strange sitting on the right side, but the pedals were in their familiar places. Not so with the three-speed gearshift on my left, about which I had real misgivings. Timidly, I started it up and among all the vibration and cacophony of the Chevy V8, I carefully located first gear with my left hand, slipped the clutch, and moved off. On the drive back to Conrad's apartment, I got accustomed to the left-hand shifting and the Ferrari and I began to become friends.

The car was so light that even though the Chevy engine was stock with one four-barrel carburetor, it had lots of power and acceleration, which I had to get used to after my dad's heavy '57 Plymouth. The biggest problem, I soon learned, was vision, because when looking back the sharply dropping curve of the top and its low rear window only gave a view of a small patch of ground about 10 feet (3 m) behind the car, and while the low-mounted rear-view mirror did give a better angle of view, it was convex and distorted perspective, plus it shook. Not only that, but the big fuel filler tube through the rear window was also blocking a large part of what you could see. Looking forward, there was the high, long hood which seemed to go on forever and ended who knew where – and there were no bumpers. On top of that, you were seated on the right, so in close traffic there was no sense of what was happening up ahead.

All of the instruments were working except the tachometer, which had no output connection from the Chevy engine. At least with the temperature and oil pressure gauges I could watch for trouble and that was comforting. The speedometer was calibrated in kilometers and was well off the mark anyway, due to the gearing from the Chevy transmission, but I began to establish reference points for speed limits. There was another instrument at the far left of the dash, something peculiar that seemed to involve a clock and a speedometer, but apparently it was disconnected and not important for my driving.

When I got to Conrad's apartment building and parked, he and his wife Cheryl had left, so I sat on his steps and stared at the Ferrari, thinking, "Look what I have."

The next day, the 6th, I set off due south and facing the sun to Mobile, wearing a T-shirt to help relieve the July heat, but that did not help much. Opening the vent windows in the doors and the forward-hinged rear quarter windows to their full extent of a few inches and opening the foot well vents only created a flow of hot air through the cabin. In those days, however, air conditioning was not the norm in cars and my dad's Plymouth did not have it, so I was somewhat hardened to the heat. At 200 miles (322 km), though, it was an uncomfortable trip, especially being uneasy with the strange old car and fearing that it might break down at any minute. I did discover that the car's unusual feature of a knee brace protruding from the driver's door was a blessing, allowing me to rest my accelerator pedal leg on it in my folded position, and it was good for a prop in turns as well, something that undoubtedly had come in handy in its era of racing without seatbelts.

I drove without stopping, fearing that the car might not start again if I did, and finally pulled into the backyard drive at my parents' corner lot house, parked under the shade of an oak tree, and got out to cool off. My mother came to the back door, looked at what I had brought home, and said, "You paid fifteen hundred dollars for that thing?"

As the photographs I took that day show **[4-10]**, the Ferrari certainly looked bad, stripped of paint with splotchy spots of filler, and a photograph showing sister car 0048M stripped of paint **[11, page 18]** gives a better idea of how it appeared in color. I knew I had a lot of work ahead of me

[ABOVE & FOLLOWING FOUR PAGES: 4-10 et al] 0046M as it appeared stripped of paint when I brought it home to Mobile. *Author*

[11] 0048M stripped of paint illustrates the motley colors of my new car. *Barchetta.cc*

to get the body smooth enough to paint and make it look presentable. Regardless that I had no experience with body work whatsoever, the next day I drove the car into my dad's garage and proceeded to learn the mysteries of Bondo with on-the-job training. About 10 weeks remained until classes began for my senior year at the University of Alabama, and if this was to be my transportation, it had to be ready by then.

It took a while to figure out the hardener/base/working-time equation, but after a few mistakes I was soon troweling filler on and sanding it down with confidence. I lacked the nerve to attempt any metal shaping, though, and the rear fenders could have used that, apparently having been leaned on enough to misshape them, so I just layered on the Bondo and sculpted curves I thought were appropriate. Fortunately, the top and hood needed filler only in a few small spots, so I mainly sat on a stool and worked along the sides of the body. The aluminum was so soft, I simply pressed out a few small dents on the tops of the front fenders with my thumbs. This was in July and August during the height of the summer on the Gulf Coast and I was working in a small wooden garage, so the heat and humidity became unbearable in the afternoon. I would begin in the morning, then work as long as I could stand it. Slowly, the car began to regain a somewhat smooth surface, although my amateur efforts would never have passed muster in a high school body shop class.

In the garage I had time to study the car, which I had not done in my haste to buy it. Now I could get under the hood and into the nooks and crannies everywhere to see how it was put together and what was there. I could see the evidence of the V12 having been in it from the two unused ignition coils still attached on either side of the forward engine bay and the unused holes in the firewall. The oil cooler was missing from the right side of the radiator. The radiator top had the central opening that the Ferrari used for water return capped off and the extension pipe that originally housed the radiator cap on an elbow was now cut shorter and used for the return from the Chevy engine.

Since the Ferrari was right-hand drive, the clutch throwout arm had been on the right, next to the clutch pedal, but the Chevy throwout arm was on the left side for left-hand drive use, so a hydraulic master cylinder and slave had been put in to actuate the Chevy clutch. For that, the firewall had been cut at the bottom by the bell housing for clearance and some patchwork aluminum sheet metal had been made up to cover the hole and the Chevy transmission. Other than that and cutting off the Ferrari motor mounts and welding new ones to the frame, there had been no need for any butchering to make the V8 fit in. It was tight, the exhaust manifold barely clearing the steering column, but it worked perfectly. At the other end of the drivetrain was a Chevy rear axle that had been cut down on each side, and the shortened Ferrari axle tubes had been slid in and welded on so that the original suspension mounts, brakes and wheel hubs could be retained. I never saw an axle shaft itself, but something must have been done so that the splines would work on each end. In all, it was actually a nice piece of work **[12]**.

On the dashboard, a modern-style key switch had been installed to take the place of the original. On a small switch panel under the center of the dash, there was a push button for the starter, pull switches for the headlights and wipers and another pull switch, which I believe must have been used for the driving lights that were no longer there. In addition, there was a manual left/off/right turn signal switch in the center of that panel. Also, there was a small manual push/pull pump mounted under the dash, slightly to the right of the passenger's grab handle, which puzzled me.

[12] The Chevy motor was a tight fit in the engine compartment. *Author*

Much later, I discovered that the car once had windshield washer nozzles and that this must have been the pump for them. From its location, most likely the co-driver would have operated it in races, but that gets ahead of the story.

Now that I owned the Ferrari, I wanted to know what it was I had. On the bill of sale Deerman prepared, he identified the car as a 166 Mille Miglia[5] and he had told me the year was 1954. Obviously it was an early car, judging from the egg crate-style grille, but to be sure I decided to go straight to the source who should know: Ferrari. I wrote to the factory on July 13 asking for information about the car and provided them with the only identification I could find, which was the number 3447 from the 23x86 mm aluminum Carrozzeria Touring[6] tag in the engine bay at the top of the firewall on the driver's side **[13]**. I was overjoyed when a reply arrived quickly, dated July 28 (Appendix 11). The number I gave them was not a part of their system, they said, but they had

[5] The Mille Miglia was a "thousand-mile" (1,609 km) race around Italy, 1927-1957.

[6] *Carrozzeria* is the Italian term for a constructor of automobile bodies. Touring was one of them.

[13] Carrozzeria Touring body number plate from the firewall. *Author*

recognized the car from it. Someone at Ferrari – a secretary, I suppose – had sat down at a typewriter in this pre-digital age and manually produced six pages of specifications in which all the parameters of the car and engine were neatly listed in tables, the first page of which is included in Appendix 12. It was serial number 0046M, a 166 Mille Miglia with its Certificate of Origin[7] dated April 6, 1950. Not only that, but they named Nuccio Bertone as the first owner and a Bruno Martignoni as the second owner. This was my first indication that my little car might be something special. I had no idea who Martignoni was, but I knew of Bertone and was quite proud that I could sit behind the same steering wheel as he had.

In my letter to Ferrari, I had asked about the Mexican Road Race and any other racing history they might know about, but they replied,

> *As a car sold to an individual customer, we were not able to follow its sporting activity. Although it is not among those employed by the factory in races in Mexico and North America, it is not impossible that it participated in some of them in those countries.*

Eager to know all I could, I turned to the only books I found available at the time and ordered Hans Tanner's *The Ferrari* (2nd edition) and his *Ferrari Owner's Handbook*. When they arrived, I looked through all the photographs but nothing matched my car. Next, I measured the wheelbase and track and those numbers matched perfectly with the 166 Mille Miglia data in the *Handbook*. Since I knew that the engine I had bought had 166 markings, it would make sense if the engine really did come from the car, but the 166 MM *berlinettas* ("little coupes") looked nothing like it and the *barchettas* ("little boats," as explained later) were roadsters. Even though I had the factory's assurance that my car was a 166 Mille Miglia, it still did not look like one, except along the sides from the horizontal crease on down, and while I pondered that I kept troweling on and sanding down the Bondo.

After many sweaty and dust-choked weeks, I finally reached a point when I thought the car was acceptable enough to paint. Time was running out, after all, and I had to get back to the university. So, at last I went downtown to a store that sold auto paint and had them mix a shade of red lacquer that looked appropriate in the can. I also bought a can of zinc chromate primer, which I had read would protect the aluminum from oxidation and help the lacquer bond to the surface. These I brought to my brother C.E. Westfaul's house where he had the compressor and spray equipment to paint the Ferrari. He was 33 at the time and had been involved with cars since he was 19, during which time he had painted many; plus, he would do mine gratis. So, I drove the car over to him one evening after he got off work and after masking it the painting started, at night, outside in the open, first with a coat of the zinc chromate, and then the next night the first of several coats of the red lacquer. Rubbing it down each day following a coat was my job, and after a few days I had a shiny red Ferrari. All that was left was to attach the adhesive Ferrari prancing horse shields I had ordered from an advertisement in *Road & Track* magazine for each side because I felt no proper Ferrari should be without those.

The car was still on the worn, mismatched set of tires visible in the photographs I took the day I arrived with it, and to keep me safe, my father offered to buy a new set of tires. Previously, I had transferred Deerman's license plate to my name by mail to the Tuscaloosa License Commissioner's office on July 12 (Appendix 10), so off we went downtown to Sears. My 76-year-old dad was in the left side passenger seat, which was where all his life he was accustomed to sitting while driving, and it was a unique experience for him,

7 A legal document with verbiage similar to that for a later car, which states Ferrari "hereby certifies that the chassis described below has been manufactured in its works," followed by serial number and data, and that "The chassis has been delivered brand new from the factory on ___."

probably a little unsettling, too, without a steering wheel. As it turned out, that was the only time he ever rode in the car.

At Sears, we picked out a set of tires and the incredulous workers set about removing the knockoff Borrani wire wheels, putting them on the tire fixture that was not made for them, pulling off the old tires and putting on the new ones with inner tubes. All the while I was trying to watch carefully to protect the wire wheels and my new paint. All done, we headed home on fresh new Allstate tires.

The next morning, I looked out at the car and saw it had a flat tire. Right away I suspected what the problem was because I had read about it in *Road & Track* but failed to think about it at Sears. The Borranis had aluminum rims, and when the tire mechanic had a wheel on the fixture and the new tire in place, he ran a steel tool around under the rim to force the bead under it. In doing that, he was peeling off sharp aluminum shavings that dropped down between the tire and the inner tube, so when the tire was inflated and rolling, the shavings cut into the tube. If one tire went flat, all the others could too, and there was no recourse but to break all of them down by hand the old-fashioned way using a piece of leaf spring as a lever, clean out all the shavings, repair the one tube that punctured, and then reassemble them all carefully by hand again. I did all that myself, inflating them with a hand pump, and there were indeed shavings in the other tires, so it was well worth the hard day's work that it took to avoid a flat on the road to Tuscaloosa.

One final job I needed to do was to get the windshield wipers working. There was a motor behind the dash that ran them but there was no linkage, so I made a rod to go from the motor to the wiper on the driver's side and then a rod from it to the wiper on the left. That homemade rod to the left side was long and kept bending and binding, jamming both wipers, so I removed it and had to be satisfied with just the right wiper. It hardly mattered, since you sat close to the right windshield and that wiper gave a barely sufficient field of vision to drive, but it was slow. From then on, I made it a point to avoid driving in the rain.

For my senior year, I had already leased a little furnished one-bedroom cottage in a group of six off an alleyway between 7th and 8th Ave., a block south of 10th St. (now Bryant Drive), which was on the south side of the campus. There were three cinder block cottages on each side of a central gravel drive, of which mine was on the left at the end farthest from the alley, and I had a pecan tree outside under which I could park the Ferrari to give it a little shelter. The first trip I made back to Tuscaloosa was in my dad's Plymouth again to carry the bulk of my things, since there was hardly room for more than me and a suitcase in the Ferrari. The following trip I made in the Ferrari and it went well, just the usual heat and noise.

I was getting comfortable driving it now and had mastered the art of being on the right side. On the open road, I re-learned the technique of passing because I had to drop back far enough to see if the road ahead of the slower car was clear, which meant allowing for more time than usual to make the pass, but the Chevy engine had plenty of power to make the light Ferrari take off like a rocket. Not that I pushed it. My enthusiasm for sports cars was never about speed but rather their aesthetics, so I never tried to "see how fast it would go" as I was often asked if I had.

With a red Ferrari parked outside my door, I thought for sure I would get lots of lookers and inquirers coming by, but that turned out not to be the case. Of course, the car was fairly hidden off the alley, and once the neighbors' curiosities were satisfied, we settled into a sequestered life. That was good because the doors had no locks and I had to leave it unattended when I went off to class. I rarely parked the car on campus for that reason and because I feared damage to the body that I had just spent weeks straightening. When I did park it there, I always left it where I could watch it. The library and the student union building were the only places on campus I would drive to, and then that would be on weekends when there were fewer people and less risk of parking dings. During the week, I would make runs to a convenience store a couple of miles (3.2 km) away to buy beer or go to a grocery store, but that was usually at night, so I was sparing in its use and quite protective of it.

In order to drive the car on campus, though, I had to register it with the university and affix a parking decal to the windshield. According to the regulations, the decal was to be placed in the lower right-hand corner of the windshield, a position that was fine for all the left-hand drive cars. For me sitting low behind the high cowl on the right, the decal would obscure part of my view, so I went to the University Police who were then in charge of parking and pleaded my case, but to no avail. I was forced to put the decal on the right side, so I put it as low and as far in the corner of the windshield as I could, and there it remains to this day **[14].**

The pecan tree at my door came in handy as my proverbial shade tree for tinkering with the car as issues developed. The starter began to be erratic, and I discovered that was due to the high current switch on the starter solenoid. But, I learned that I could work around that by opening the hood, reaching down, and shorting across the terminals with a screwdriver to start the engine in neutral. Other electrical problems cropped up, and I would spend time tracing wiring and making do.

Once, I had removed the taillights as part of a project, and at the end of the day I decided to go for a test drive. Along the way, I went by Conrad's apartment and picked him up. We were headed out on University Boulevard toward the edge of town at dusk when I saw the flashing lights of a police car behind us, so I pulled over. The officer got out and walked up to the car's left side where Conrad was sitting and Conrad looked up at him; whereupon, I got out on the right to talk to him since the windows were fixed in place. When I did that, it dawned on him that there was no steering wheel on the left side and it was probably his first time seeing that, so he had a moment of confusion. I asked him why he had stopped me and, gathering his composure, he walked to the back of the Ferrari and shined his flashlight on the empty holes where I had removed the taillights. "Did you know you don't have any taillights?" he said, asking the obvious to assert some authority. I told him I did, that I had been working on the car, and that we were out for a test drive. Then the officer started asking questions about the car, so we had a nice conversation. In the end, he simply told me to get the lights working and sent me on my way, never even asking to see my license.

[14] My University of Alabama parking decal for 1966-67. *Author*

Fall was tag renewal time in Alabama then, and on October 11 I drove down to the Tuscaloosa County courthouse and renewed the license for my car, this time in person. There was less surprise over the make and model than I expected, probably because as a university town Tuscaloosa accumulated many foreign automobiles and it was not uncommon to see them on the streets, as I had seen mine. I recall one Mercedes 300SL Gullwing that was simply kept parked by the curb. My registrations are in Appendix 10.

When the Thanksgiving break came around, I headed back to Mobile in the Ferrari. I was confident enough of the car by this time to stop for gas in a small town. This was back in the era of full-service gas stations, and a guy younger than I was came out to pump gas. For him, it was as though a spaceship had landed. I opened the quick-release gas cap for him and he was amazed at the huge opening, peering down into the tank. (Incidentally, when I worked on the car and needed some gas to clean parts, I would reach my arm down into the tank with a cup and scoop some up.) The attendant wanted to know how much it held and I told him 30 gallons, which was only a guess. Astonished, he asked if I wanted to fill it up, to which I had to disappoint him. When he finished, he walked around the Ferrari, looking it over, and stuck his head in the open driver's door. He looked at the speedometer and its calibration in kilometers and excitedly questioned, "This car will go 240 miles an hour?" Sadly, I had to disillusion him, but I am sure I left him with an often-told tale about the day a Ferrari stopped for gas.

The return from my next trip home, for Christmas break 1966, was not so uneventful. I had made it down to Mobile without incident, but the day I had to drive back in January 1967 was windy, chilly and overcast with rain threatening, which was unwelcome with my one slow windshield wiper. It was a cold drive too, since there was no heater and plenty of leaks to bring in the winter air. I made good progress – until I reached the top of a little railroad bridge in the town of Demopolis, sixty miles (97 km) from Tuscaloosa, and there the engine simply shut off. I could not get it to fire at all, so I rolled backward down the bridge and into the parking lot of a business right at the base. That was easy because it was Sunday and there was no traffic, but it also meant nothing would be open to help me.

I found a phone booth nearby, and in the directory book they usually had in those days I saw a service station listed, which I recognized as being on the highway a few miles back. I called the station and to my great fortune they were open and had a wrecker. I knew better than to tell the guy what I was driving, so I simply explained that my engine quit and asked if he could tow me there and look at it. He agreed, but when he arrived and saw the Ferrari he started backing out of helping me. However, I assured him that it had a Chevy engine in and that he might indeed be able to fix it. So, we attached a tow chain as I would not allow him to pick up my car with his hook, and I was pulled along behind, cautiously braking when he slowed or stopped.

His service bay was full, so we had to park the Ferrari outside in the breezy, drizzling cold where I opened the hood for him and he began to poke around. The hood on 0046M is nearly five feet (1.5m) long and about three feet (0.91m) wide at the cowl end; plus, it is curved convexly and weighs very little, so it is much like a sail. The prop is a simple little rod that slots into a hole in the hood's frame, and it is literally just a prop, without a catch. As I was standing beside the mechanic, a gust of wind caught the hood and blew it forward, off the prop, and before I could rush around to catch it, it went vertical and the forward edge hit the hinge supports, gouging a mark on the left side and one not so bad on the right. I got in front, grabbed the hood in a bear hug, and stood there fighting the wind from doing more damage. I had to remain like that as the mechanic worked, holding the hood that was much taller than I was and straining against gust after gust in the cold, but I was thankful for the help and that the man persisted under those conditions.

Eventually, he found that the button that contacts the rotor inside the distributor cap had simply disappeared without a trace. He was able to get someone he knew to open a parts house and meet him while I waited with the hood closed but damage done to my new body and paint work. Once the new distributor cap was in, the engine fired right up and all that remained was to pay the man. I have no recollection of the bill, but in a time before credit cards, I did not have enough cash. Like it or not, he took all I had and relied on my honesty to mail him the rest, which I did. Finally, I was on my way again.

When I was doing my bodywork on the car, I had repaired a mark on the leading edge of the hood exactly like the one it acquired that day, so the hood either being blown over by wind or overextended by mistake had happened before. Early photographs that follow will reveal that mark on the left side.

Through the rest of the winter I froze while making my occasional beer or grocery runs; otherwise, the Ferrari mostly sat outside my cabin. I was close enough to walk to classes on campus, and with no social life to speak of, I never went on dates with the car, not that it would have been particularly impressive anyway with its lack of creature comfort.

Warmer weather brought spring break and once again I left for Mobile, this time with the prospect of a more comfortable trip. However, a few miles outside town I was driving along when the car suddenly started slowing even though the engine was still running. I pulled to the side of the road and discovered that the transmission was in neutral. So, I shifted back into first and moved off, but when I got into third gear, it eventually jumped out again. My only recourse was to drive the full 200 miles (322 km) holding the shift lever in third. When I got home, I talked to my brother about it and he explained that something was probably worn out in the transmission from drag racing. He arranged for me to take the car to a friend's shop where they pulled and rebuilt the transmission as a rush job so I could get back to classes. That was on Saturday, March 25, 1967, and Hay's Garage on Springhill Avenue charged $60 for the work.[8] After that, I had an easy trip back to Tuscaloosa.

It was not long after that I noticed the clutch was beginning to slip, which was probably another sign of previous hard use in drag racing. When a nice weekend came around I set up shop under my pecan tree, took out the seats, removed the transmission cover, put a jack under the rear of the engine, and with Conrad's help, pulled out the transmission. I had watched the mechanic do it when the transmission came out for repair, so that much I knew about. Working on the clutch was new for me, though, but we got it out and sure enough, the friction plate was worn thin. So, we went to a parts shop, bought a new friction plate, and then put the clutch back together on the engine. The problem I foresaw was aligning the friction plate to allow the splined transmission shaft to mate up and slide through, since we had no tool to do that, but we got lucky and managed to wrestle the transmission back into place.

The rest of my senior year passed without incident and I graduated in late May, returning in the Ferrari to Mobile and ferrying the remainder of my things back in the Plymouth. I had driven around 2,500 miles (4,023 km) in the Ferrari, and every one of them had been an adventure, but now I saw the need for more creature comfort. I had worked on the exterior, but the interior needed to be cleaned up, sealed and carpeted. To do that, one day in June, 1967 I drove it back into the garage, never imagining that would be the last time I would ever drive the car.

[8] In 2024 dollars: $570.

With the Ferrari in the old garage, I began stripping the interior, removed the seats, and found the Touring body number 3447 written in blue crayon on the elastic webbing under each one, although at the time I did not know what it was **[15-17]**. To have full access to the back of the car, I removed the fuel tank and its surrounding panels. Removing the tank took quite a bit of trial and error since it is huge and almost completely fills the space behind the rear axle. With the seats out of the way, I managed to inch it up over the rear axle and out through a door.

I removed the headliner, beneath which could be plainly seen the hammer marks in the bare aluminum from when the panel for the top was beaten into shape. I removed the steering wheel and dash panel too, carefully boxing up the instruments, labeling the wires that had led to them and drawing a diagram. It was that storing of the instruments which inadvertently led to their preservation in pristine condition for all the years that were to follow. With the dash removed, I could reach the firewall better and I stripped away the insulating layer of jute mat that had been glued to it over red paint. Some of that jute material I began to reuse as a filler, stuffing it into the gap between the body and the top of the inner fender panels where there was quite a draft and cementing it in place with roofing tar.

As the project dragged on through the heat of the summer, it became apparent that I would not be able to get the car back on the road in time to return to the university with it. I was going back because I had been accepted into graduate school with the objective of a PhD in English and had been awarded a graduate assistantship teaching freshman courses. For substitute transportation, my sister Fay Wainwright offered the loan of her 1958 Chevy Bel Air, so there was no rush to work on my car. And thus the long sleep of the Ferrari began.

My first year of graduate school ended and I returned home for the summer of 1968 and returned my sister's Chevy, leaving me without a running car of my own except my dad's Plymouth. I puttered around with the Ferrari but there was no end in sight, so as time to return to the university

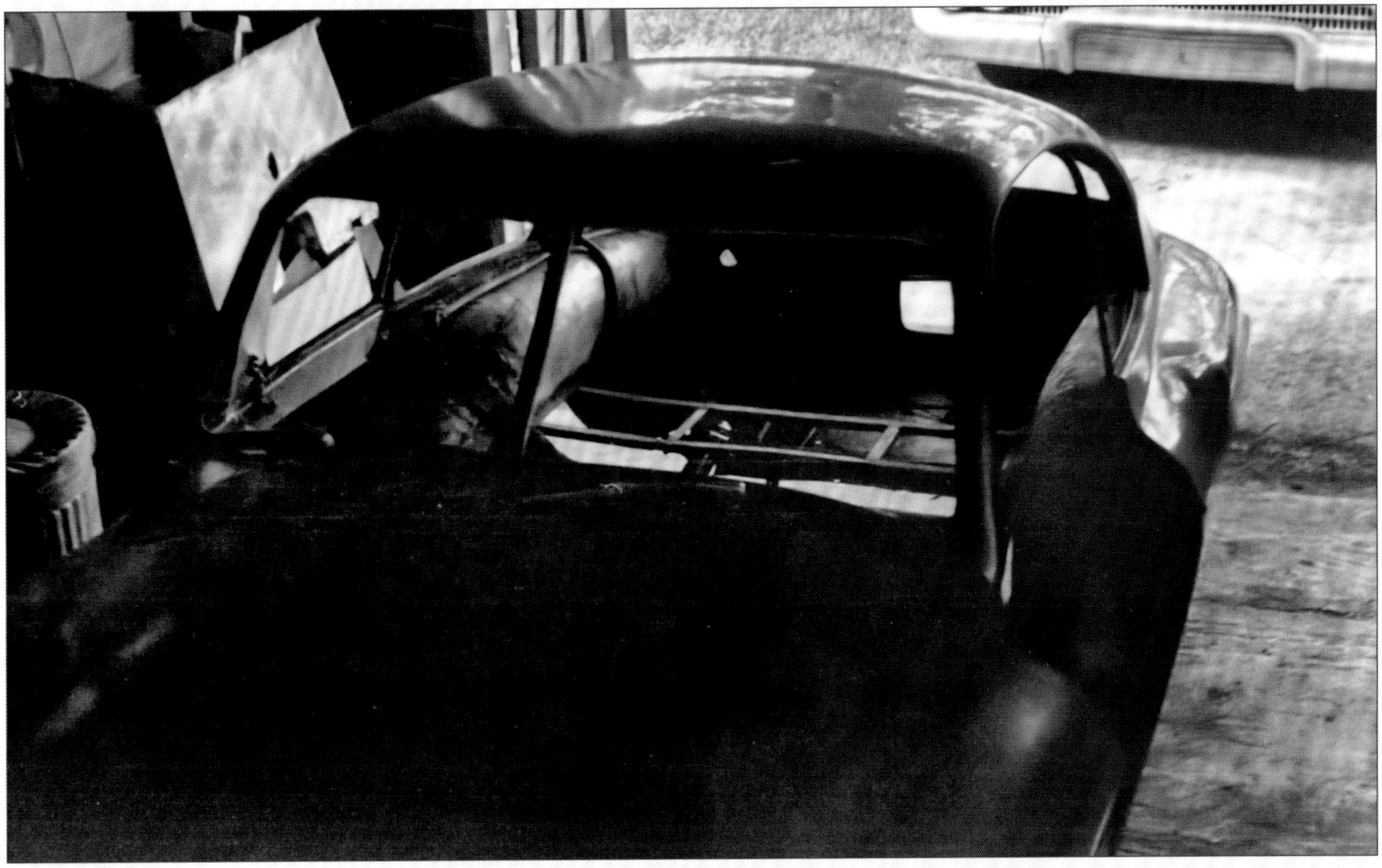

[THIS PAGE & OVERLEAF TOP: 15-17] Disassembly began in the summer 1967 for interior refurbishing. *Author*

began to grow near, I bought a 1962 Jaguar E-Type roadster with the help of my brother, paying $1,700 for it.[9] With that, I had a much more practical car, and while it had its quirks, it did have a heater, locking doors and windows that rolled up and down. It needed tires, though, and conveniently the new tires on the Ferrari would fit, so I made the swap and the Ferrari acquired the worn tires on which it stayed for decades. In September, off I went back to graduate school in the Jaguar, my second car, which I still have.

In subsequent summers, I began teaching in summer school at the university, so the time that I spent in the garage with the Ferrari became less and less. Then, in the spring of 1971, my father died and my mother decided to sell the house and garage and move to my sister's home on Mobile Bay. So, after packing up the loose parts and putting them inside the Ferrari, my brother and I loaded the car and the boxed engine on his trailer for the trip. We positioned the car along the blank north side of the house, out of the way, and wrapped it in a canvas tarp **[18]**. The engine parts were stacked on the long front porch in their crates. There was no possibility of doing any work on the car at that point, and I was getting back to Mobile less, so the Ferrari became an afterthought.

After a year, my mother missed being in the city, so she bought a small house back in Mobile with a somewhat dilapidated garage in the back yard. Again, my brother and I loaded the Ferrari and parts on his trailer and hauled them back to the city. In the short space between the house

[18] Stored under a tarp at my sister's house 1971-1972. *Author*

[9] In 2024 dollars: $15,870.

and garage, it took some time to align the trailer with the small double-door opening, but finally the Ferrari rolled off and into its new home with the engine parts stacked behind it. Over the years that it stayed there, the garage continued to deteriorate around it. The floor under the car was wood planks on joists, but over time they either rotted or the weight of the car caused the floor to collapse under the wheels and the car sank. It served as good protection, though, and remarkably survived intact after the direct hit of Hurricane Frederic in 1979, narrowly being missed by a falling tree. It also provided what is surely one of the all-time best barn find photographs **[19]**.

As the years went by, I began to realize that the Ferrari must be increasing in value. I recall seeing an advertisement in *Road & Track* magazine in the early '80s with a 166 MM *barchetta* for sale at a price of $82,000.[10] While I was still not clear on what my car really was, I had the factory's assurance that it was a 166 MM, so the rising tide must have been lifting my little boat too.

[19] Stored in a dilapidated garage at my mother's house 1972-1988. *Author*

My life had undergone changes as well. While in graduate school and also teaching, I had started a sideline business with a few pinball machines in 1974. They were placed at locations on a profit-sharing basis, and with the popularity of pinball at the time, profits began pouring in. Reinvesting, I bought more, diversified my equipment, and expanded my locations until I saw that my business could be far more lucrative than a career as an English professor. By then I had married my wife Elizabeth and financial necessity outweighed idealism, so even though I had finished all of the coursework for my PhD, I walked away from the academic life in 1976 and went into the amusement business full time as Dixie Novelty Company.

All along I had wanted to trace my Ferrari's history and discover what it was and where it came from, but I had hardly any spare time due to the demands of my growing business. That changed on New Year's Eve 1987 when I had a wreck and fractured my right wrist, sustaining other more minor injuries to my legs as well. Unable to use my right hand and hobbling around, I was forced to delegate daily operations to my employees while I stayed home to recuperate. It was then that I decided to make use of my down time to seriously look into the history of the Ferrari, and since I was becoming proficient with my left hand, I could use the telephone and my electric typewriter as my tools.

[10] In 2024 dollars: $270,000.

Chapter 2

THE TRANSMISSION

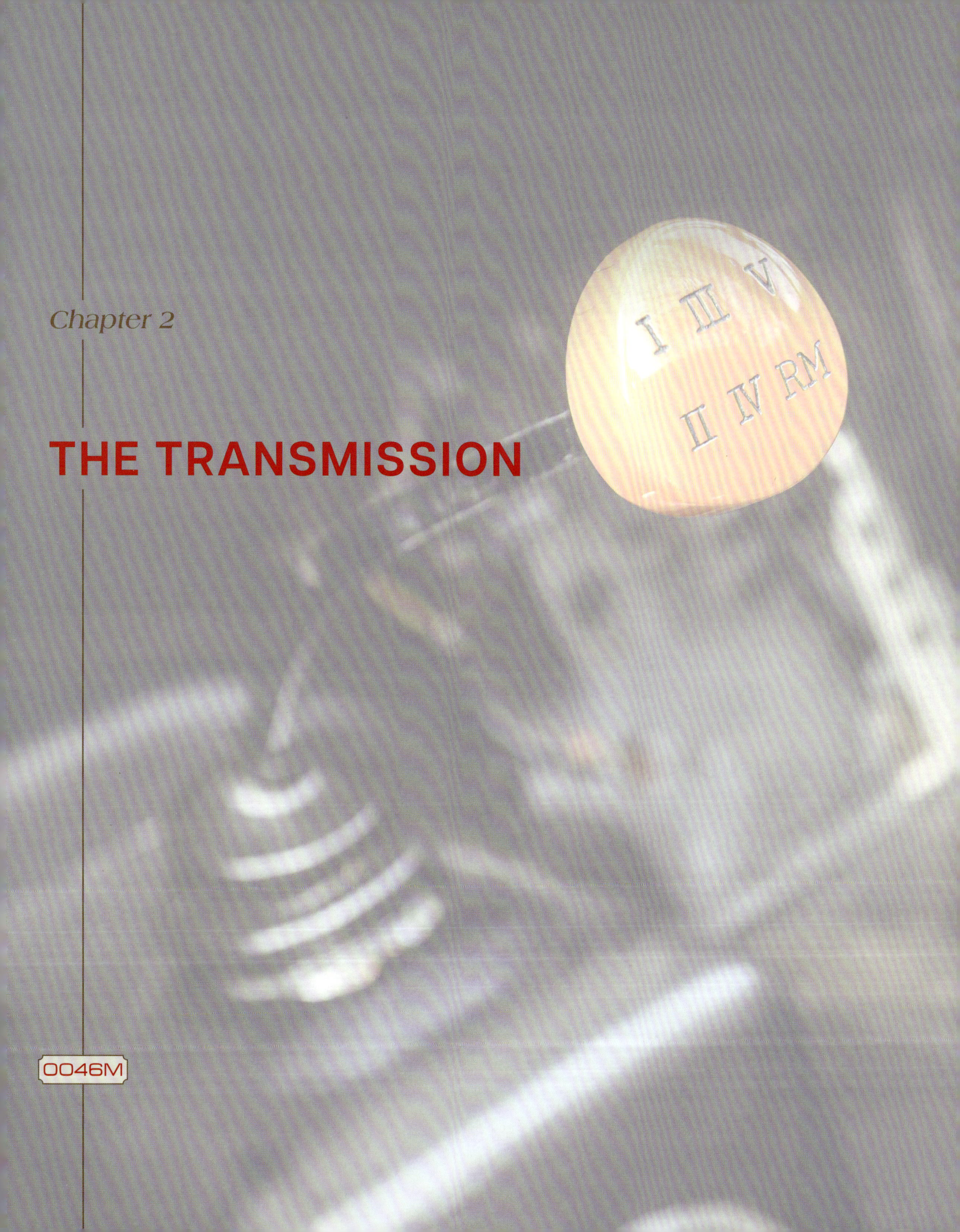

0046M

I had two objectives as I began my convalescence: to learn the history of the car and to find the original transmission. As a 21-year-old, my appreciation for originality was not as keen as it was now at 43, and if at all possible I wanted to reunite the engine and transmission. Even if the bell housing had been cut off, I reasoned, something could be done to fabricate an adapter or replacement. So, I began hunting the Lotus 11 with a Corvair engine in front that had my transmission, but this was long before internet search engines and after 22 years the trail was now very cold.

I had seen a newspaper article about David Whiteside, a Birmingham man who vintage raced a Lotus 11, so with my good left hand I dialed Information, which in those days had everybody's number, and called him. He had never heard of such a car and frankly doubted such an engine swap could be done, but he gave me the names of others I could call and the number of Bruce Moritz, who was president of a Lotus owners club. Bruce was willing to help me when I got in touch with him, but for privacy concerns he would not release his list of Lotus 11 owners, so we compromised. Again with my left hand but now on the typewriter, I composed a form letter and had a local copy store run off 50 copies. These I sent in envelopes with postage to Bruce and he affixed address labels and mailed them. In time, replies began to trickle in. They were all sympathetic but no one had ever heard of such a car or one where a Corvair engine had been removed.

After weeks of futility, following leads and making cold calls, it seemed as though I faced a dead end on finding the Lotus 11. One evening I called Bob Rabun to see if he had any clues on where the car might be. Bob was happy to hear from me after two decades, but unfortunately, he had no idea what happened to the Lotus. Even so, we did have a productive conversation as he filled in more details about the engine and also revealed what I had not thought to ask years before: the engine came from a Maurice Blevins in Birmingham, he said, and it had indeed come from my car, which Blevins had owned.

The name of Bob's partner, I learned, was Michael (Mike) Ken Carpenter, and the engine was intact when they had picked it up from Blevins' residence at night in a truck. They took it to Mike's basement garage in Birmingham at 8323 12th Ave. South where they wired up a battery, eager to see their prize turn over, but it was locked up solid, something they were not told when they bought it. Thus began their efforts to repair and rebuild the motor, leading to their eventual capitulation and the migration of the transmission to the Lotus. But Bob had lost touch with Mike and did not know where he was. He was commiserating with me when he mentioned that Mike had simply vanished after he graduated from the University of Alabama. That was it. As soon as I ended the call, I literally turned around, pulled my copy of the *Alabama Alumni Directory* off my bookshelves, and within five minutes I had Mike's phone number and was talking to him.

Once I told him of our connection through the Ferrari engine, Mike was quite pleased and we talked for a long time as he shared his part of the story. When I asked him the whereabouts of the Lotus 11, he told me that in the late '60s two young guys saw it in his yard in Birmingham, asked about buying it, and he sold it to them. That, sadly, was the last he knew of it. He did tell me the story behind the car, though, which was interesting. He had wanted to use the Corvair engine for its power, light weight, and low profile in the Lotus, but that engine was designed to be located behind the rear axle, which meant that if it were front-mounted it would be turning in the wrong direction. His solution was to consult with Zora Arkus-Duntov, the "Father of the Corvette," at Chevrolet, who thought the project novel enough to help. Together they got the Corvair engine to run backward, and it worked.

While he was no help with the transmission, what Mike told me about the Ferrari engine was enlightening. He was the mechanically inclined one of the partnership and had overseen the attempts at repair while Bob Rabun mainly provided the financing, hence the reason why the engine had remained untouched under Bob's bed. As they set about trying to repair the engine, the problems centered on the cylinder that had dropped a valve due to the breaking of a valve spring. Mike welded in some aluminum filler where a small hole had been gouged in the head. They had the valve seats replaced, found valves that would work and had them machined to fit. Figure **[20]** shows the damaged combustion chamber and its repairs compared with the adjacent combustion chamber while **[21]** shows their relative size. They replaced the cylinder liner by having a substitute machined down to the right size and shape then pressed into the block. They found a replacement piston. The rod was bent and their approach to that was unique: They located someone with arcane skills and through a long process of heating, tapping and tweaking, he actually got the rod straight again. The final piece in the puzzle was the mousetrap-style valve spring. They looked at similar motorcycle engine springs, but none would do. Some thought was given to having them custom made, but in the end they gave up and for want of the spring the project was shelved and the engine and transmission parted ways.[11]

[11] Why they did not order the spring from a Ferrari dealer, he never explained. Perhaps they were sold in sets and were too expensive.

In retrospect, it was good that they were not successful since more damage might have resulted from an inexpert assembly of the engine when it was started. In hiding under Bob's bed, it had lived to run another day.

Faced with the likelihood that the Lotus 11 with my transmission might never be found, I joined the Ferrari Club of America in the hope of networking to locate a substitute.

[20] Damaged cylinder head chamber number 10 is on the left with amateur repairs. *Author*

[21] Combustion chamber size compared to my hand. *Author*

I kept an eye on *Hemmings* magazine as well, in case one should be advertised for sale there. What did turn up in *Hemmings* was an advertisement for a 166 engine for sale. Seeing that, I thought it would be good to have another engine for spare parts, so I called the number listed. It was Cooper Weeks. He said there was no transmission with the engine and when I asked what he wanted for it, he said, "Sixty." That confused me because sixty dollars made no sense, so I questioned, "Thousand?" and Cooper said yes.[12] I was completely oblivious to the fact that Ferraris had become blue chip investments and that their prices and that of their parts were now rising exponentially. The spare engine was beyond my means, so I declined, but Cooper turned out to be the owner of a 166 Mille Miglia himself, 0022M, and with Ed Gilbertson, who owned 0058M, was organizing the Fortieth Anniversary Reunion of the Touring *barchettas* and owners for the next year, 1989, in California from August 14 to 21. He invited me to come, even though my car could not, and sent me an entry form which I happily submitted. I felt like I was becoming a part of a fraternity, an exclusive one.

The conversation with Cooper also proved to be educational. When I gave him my car's serial number, he asked if the motor number matched, but I had never given a thought about the motor having a serial number. Embarrassed at my ignorance, I asked him where to find it and he told me to look on the right rear motor mount. So, on my next trip to my mother's house in Mobile, I paid a visit to the Ferrari in the garage and pried open the crate that contained the block. Sure enough, there in the beam of my flashlight was the number M 0046 **[22]**, and on the side was the European format date 2-1-50 (January 2, 1950). Needless to say, I was elated to know for certain that I really had the right engine.

Still looking for a transmission, though, I chased leads on possible sources. I called Dick Merritt, who had some spare engines and transmissions but declined splitting them up. I tried Chuck Betz and Fred Peters, who had accumulated a store of parts and were restoring several old Ferraris, but they had no 166 transmissions. They were friendly but discouraging in telling me that what I needed would be nearly impossible to find and even if I did it would be very expensive.

I did learn of one early transmission that became available. Richard Freshman at Fossil Motorsports in California had one that had come from a car that caught fire and burned at some point. The case had melted but I was told that the gears were salvageable. The price: $26,000.[13]

By early 1989, I had begun to lose hope. Once I had regained the use of my right hand I had become quite active in the Ferrari Club of America, though, and organized the first of my several track events that spring. This allowed me to get to know the Southeast Region members, one of whom was Dr. Mark Clark in Birmingham. He had recently bought a project car, a 195 Ghia Ferrari that came with a dismantled engine but no gearbox. He invited me to come see the car, and as we looked it over, he said he should be able to find a transmission for it easily. I begged to differ, but he was quite confident.

A month or two passed and Dr. Clark called me about some club business. When I asked how his hunt for a transmission was going, he had to admit I was right: he had had no luck at all. He did say that the only lead he

[12] In 2024 dollars: $160,000.

[13] In 2024 dollars: $66,670.

[22] Engine serial number. *Marcel Massini*

had gotten was that John Bond, the deceased former publisher of *Road & Track* magazine, had owned a spare engine and transmission with his two 166 MMs.[14] I sympathized with him, declared that it was a hopeless effort, then got off the phone with him and called the office of *Road & Track*. I found John Bond's son, John, Jr., who told me that indeed his father had owned a spare engine and transmission, but that they had been sold with the two 166s to Tohru Horinouchi in Japan. Unwilling to let the trail end there, I asked Bond how to contact him. He did not have that information, but he did recall that Tohru was a member of the Gullwing Club in the United States, for which he had the phone number of Lynn Yenkyel, the president. So, I called him. Yenkyel was at first most unwilling to divulge any private information about a member, which I understood, but I explained my situation and that this might be the chance to find the missing part that would complete a rare car. As one of the brotherhood of thoroughbred car owners, he did the honorable thing and gave me the phone number and address of Tohru's business in Japan. Fortunately, I was told Tohru had graduated from UCLA and spoke English well.

In case I was unable to reach him by phone, I wrote Tohru a letter on April 19 explaining who I was, what I had, and what I needed. Soon after, I telephoned, timing my call to be received mid-morning in Japan. Someone answered in Japanese, and I asked if they spoke English; they replied again in Japanese, so I simply said, "Tohru Horinouchi." There was more Japanese and I repeated his name again. Another person came to the phone and said something in Japanese, and again I said, "Tohru Horinouchi." They said something unintelligible and then there was silence. After a few minutes, another voice came on the line, saying, "This is Tohru Horinouchi."

Doing my best to be polite but earnest, I explained that like him I had a 166 MM and that it was my dream to restore it to its original condition with its original engine, but that I had exhausted all means of finding a correct transmission. Since he had a spare engine and transmission, would he consider selling me his?

Tohru replied that he did indeed have the spare engine and transmission, but that he had intended to use them as a display. He said he would think about it, though, and I told him I had a letter on the way which would affirm my goals, to which he said he would give me a reply. When I asked him if he knew about the upcoming *barchetta* reunion in August, he replied that he did but had not yet decided on going. I told him I would be there and we ended the call on a cordial note.

I was ecstatic. Here at last was a chance, although it was by no means a certainty, but I had a good feeling about it from our conversation. Tohru had not said no and he seemed concerned about my plight, so maybe this could work. Weeks turned into months, but I still had no letter from Tohru. Finally, I received a letter dated July 3, in which he wrote the following:

> *First excuse me for my long delinquency in replying to your letter dated April 19, 1989. It was nice to talk to you over the phone when you called me.*
>
> *I have decided to come to the event to celebrate the 40th Anniversary Reunion of Ferrari 166 Barchettas, as you might know it through Mr. Weeks. So I am looking forward to seeing you at the place.*
>
> *As regards your inquiry for my spare transmission, to be honest, I would like to keep it at my place, but as a fellow lover of the 166 barchetta car, and in the belief that I should assist you in your wish to restore your barchetta as authentically as possible, I will sell that transmission to you.*
>
> *I would like to talk about it when I see you at the event. Looking forward to seeing you soon, I remain*
>
> *Sincerely,*
> *Tohru Horinouchi*

Success! I had the promise of a transmission but now I had another concern: Tohru would sell me his transmission, breaking up an exceedingly rare unit, but what would it cost me? I knew that its value would be well into five figures. My

[14] 0020M, a *barchetta*, and 0066M, a *berlinetta*. The spare engine was 0036M.

business was successful but growing and required continual capital reinvestment, so while the cash flow was good, my personal income was only modest. I scraped together what I could and came up with $5,000, which I expected could only be a down payment and I would have to borrow the rest.[15] All I could do was wait and see what developed at the reunion.

Of course, I said nothing to Dr. Clark about what I had done with his tip. He eventually gave up on his 195 project and sold it.

I met Tohru and his wife Akemi when the group gathered for dinner the first evening of the reunion in Napa Valley at Yountville, California, on August 14. He had brought with him Simon Harrison from the Los Angeles office of his business, who had seen to the shipping and delivery of his 166 *barchetta*, number 0020M. Harrison spoke Japanese and served in a multipurpose role as an interpreter and an assistant for them. I never saw how much luggage they brought, but I recall that both Tohru and Akemi dressed impeccably, with him wearing a different suit to dinner each evening over the seven days **[23]**. This was a time of a robust economy in Japan and Tohru had become quite wealthy, enough so to amass a collection of cars that included not only the two 166 MMs, but also Mercedes 300SLs and several other highly coveted Ferraris, one of which was a 250 GTO, the Holy Grail of Ferraris.

[15] In 2024 dollars: $12,720.

[23] Tohru (left) and Akemi Horinouchi with their 166 MM, 0020M, at the Barchetta Fortieth Anniversary Reunion in Napa Valley, California. *Author*

During that first dinner, Tohru and his wife were friendly but reserved and taking it all in, probably to get the measure of the strangers and their surroundings. He did speak English fairly well, although he seemed to prefer listening and was not a garrulous man, as a few of our American company were. We sat near each other – I saw to that – and talked around the subject, not speaking of the transmission but just sizing one another up through small talk. Toward the end of the evening, he asked me if I was a teacher, to which I replied I had been in the past, and I took that as a good sign, that he was regarding me as a serious and respectable person.

Soon after, Ed Gilbertson was asking about the state of my car when I mentioned to him that I was expecting to buy the transmission from Tohru's spare engine. He warned me that Damiano Amoruso from Milan was coming, hoping to buy both Tohru's engine and transmission for 166 MM number 0036M, from which they had originally come. That worried me. I had learned that Tohru had a sense of historical propriety and that was the factor I was relying upon for him to make my car whole again, but here was someone with a car having a greater need which was in his power to fill. After meeting Amoruso, however, he struck me as having a personality completely different from Tohru's and he seemed more interested in the increased value of 0036M with its correct engine than with the significance of making the car original again.

With the rapid appreciation of Ferraris, people were buying old cars, restoring and then flipping them like others did houses, but with far greater profit. I had stressed to Tohru in my letter and in our one face to face talk during the reunion about the transmission (which I had waited patiently for him to initiate) that I wanted it for the sole purpose of restoring my Ferrari and keeping it as I had already done for 23 years and not to sell it for profit. Even so, I still had a concern over whether Tohru would change his mind.

The reunion lasted seven days, involving eight 166 Mille Miglias,[16] and while I thoroughly enjoyed the events, the cars, and meeting the other owners, and I had accomplished other objectives that will be seen later, Tohru's transmission overshadowed all else. On the last night of the reunion, in Monterey after the Pebble Beach Concours, my wife and I were in our hotel room when the phone rang. It was Simon Harrison and he wanted to come see me, so I knew this was the moment I had waited for.

Harrison arrived and began a speech that went something like this: "Mr. Horinouchi has decided to sell you his transmission. He is doing this because he wishes to help you complete your car and believes that it will be going to a good home. In exchange for it, he wishes you to pay $5,000. You understand that this is just a token amount and that it is worth far more."

[16] 0006M, 0008M, 0010M, 0020M, 0022M, 0050M, 0058M, 0068M.

It was a miracle. I was going to get an unobtainable transmission for exactly the amount of money that I had available to pay for it. I assured Harrison that I certainly understood and appreciated that such a low price was pure generosity from Tohru. I had the opportunity to thank him in person the following day before we all went our separate ways on the 21st.

Afterward, Tohru and I exchanged magazine articles from our respective countries about the reunion. As Christmas approached, I sent Tohru and Akemi a Christmas card and they sent an exquisite Japanese version in return. That began a tradition with us and we have exchanged cards at Christmas time ever since, for 36 years now.

As friendly as we were, much time passed and I was getting anxious. Finally, the following year a letter came from Tohru dated October 18, 1990, saying that he was ready to complete the sale and ship the transmission. Immediately, I mailed him an international bank draft for the $5,000 and gave him the shipping instructions.

At last, I picked up my prize on November 21 from Customs at the Birmingham airport. It was a wood box with the inside lined in sheet metal. Nestled in that was the somewhat oily transmission, which nonetheless was a beautiful sight to see, and a shifter top which I immediately recognized as the wrong kind. It was an extension shifter for a 212 Ferrari with linkage rather than the simple lever acting directly on the gear shifting fingers that I needed. In addition, there was no clutch throwout arm for the bell housing. The transmission itself could not be faulted and its serial number was 07, remarkably the previous one in sequence to the original's serial number of 8 from the Assembly Sheets **[24-27].**

I wrote to Tohru, letting him know of the safe arrival and asking if he had the correct shifter top and the throwout arm. He replied that he had checked again and found

[24] I happily sit beside my unobtainable transmission. *Author*

[25-26] The transmission and its serial number. *Author*

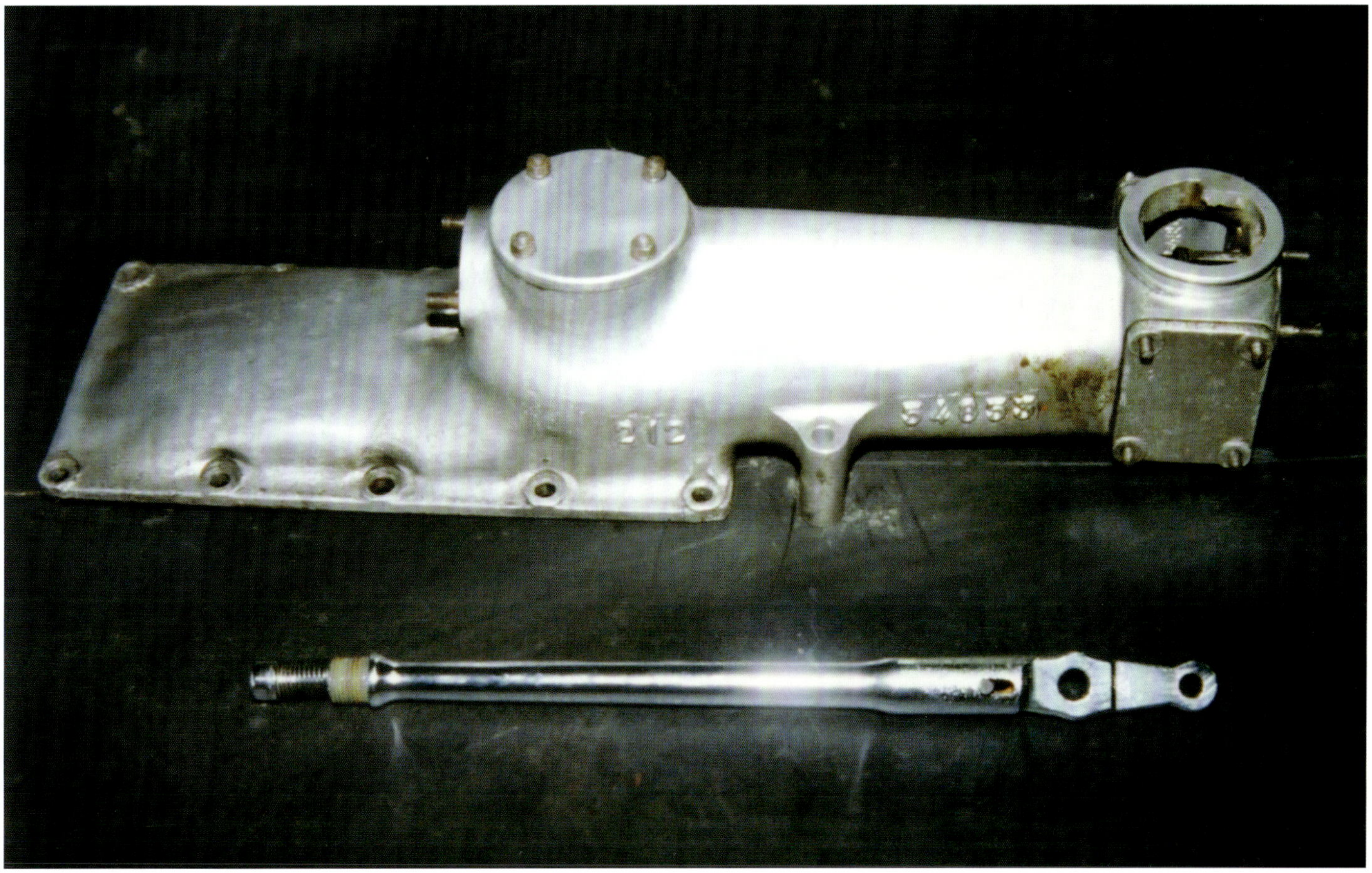

[27] Incorrect 212 extension top and shift lever. *Author*

nothing more, so I was content with what I had and was certainly grateful. I thought of either swapping with someone for a correct top, or failing that, having one either cast or machined since it was not a complicated thing and the shifting mechanism could be transferred from the top I had. I looked into those options, but none of that ever came to pass, and the transmission took its place in its box with the engine parts in my warehouse.

Before the transmission ever got to Tohru, though, it had an interesting history which can be found in *Road & Track* for February 1973.[17] John Bond wrote that his 166 MM *barchetta* had grown badly in need of an engine rebuild and that around 1960 he had purchased the spare 0036M engine (also with bent valves) and transmission as a source of parts, paying $350. So the $375 I paid for my 0046M engine in 1965 had been in line with the current market value.

[17] pp. 64-66.

Chapter 3

REVELATIONS

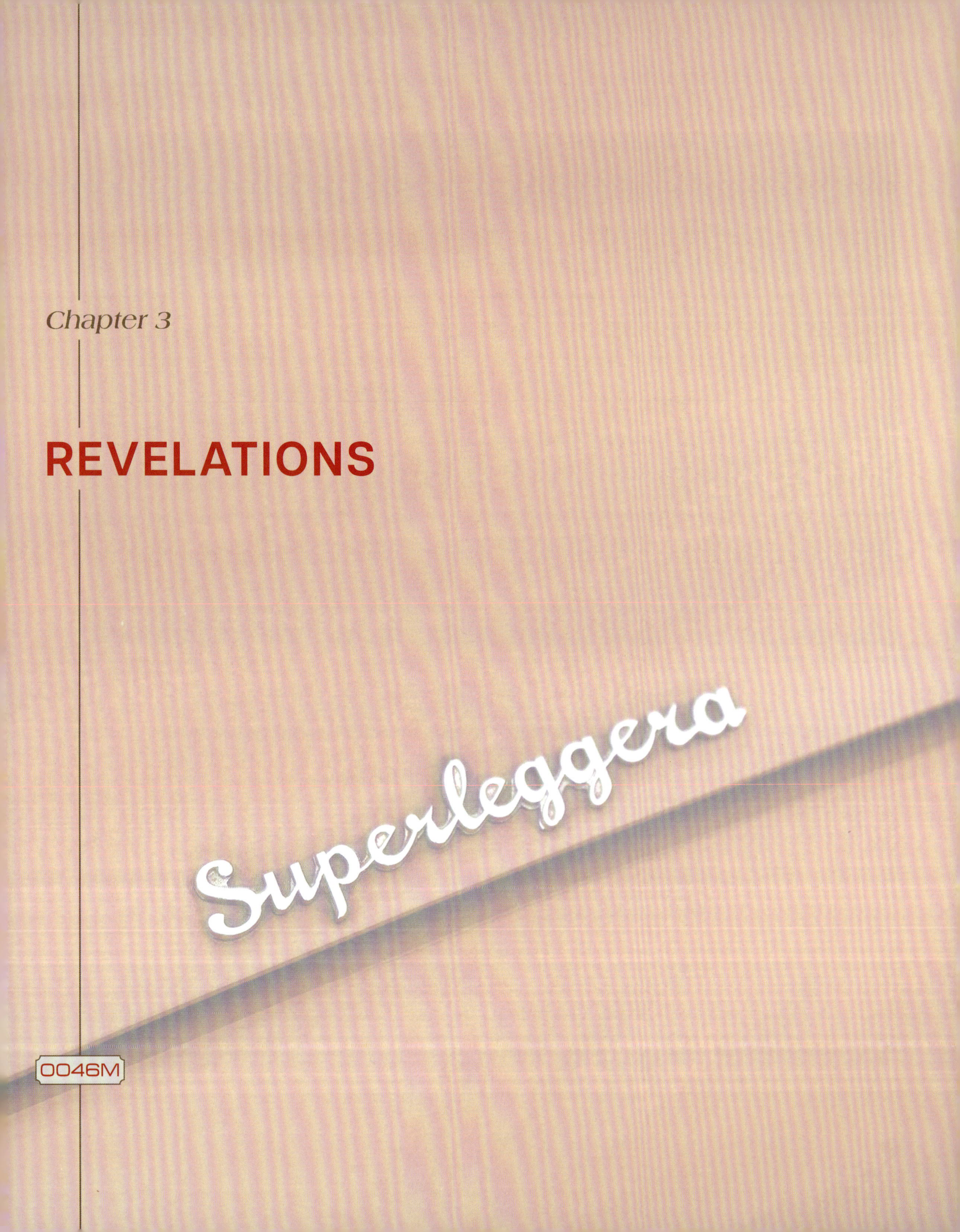

0046M

Earlier in 1987, before my accident, I had ordered a copy of Tito Anselmi and Lorenzo Boscarelli's book, *Tipo 166: The Original Sports Ferrari*. When it arrived, I eagerly thumbed through it to find the appendix cataloging all of the 166 MM Ferraris. Listed there under the serial number 0046M was the solution to the riddle of what my car was:

> *Converted to a berlinetta by Zagato in 1952-53, preserving the line of the Touring barchetta in the lower part of the body.*[18]

That explained why it met all the specifications of a 166 MM Ferrari *barchetta* but did not look like one. I had never considered that the car originally could have been a roadster. It was an epiphany after 20 years of mystery, but the book raised another question: it listed the first owner as Emilio Giletti, contrary to what I had been told by Ferrari that it was Nuccio Bertone. I also saw in the book's table of competition results that Giletti had run a number of successful races in 1952 with 0046M, showing two firsts in class and one first overall, and in the text there was mention of a third place finish. From the beginning, I was sure it must have had a racing history and this confirmed my little car had some importance, which made me eager to uncover its past. When I was sidelined by my accident, I began trying to do just that.

One of the first things I did in my convalescence was to order a book that had recently been published, *La Ferrari alla Mille Miglia* by Giannino Marzotto. When I scanned the tables of Mille Miglia Ferrari entrants, there was Emilio Giletti in 1952 and importantly the serial number of his car was listed – 0046M. Although it did not finish, here was another race for my car, so the list was growing. As I paged through the book, looking at pictures mainly since the text was in Italian, I found a striking full page photograph on page 55. It was Nuccio Bertone driving a 166 MM *barchetta* in the 1951 Mille Miglia. The serial number of the Ferrari was not given, but there on the front of the hood were those two familiar slits my car had that I had not seen in the photographs of any other 166 MM. So, more than likely this was my car when it was a roadster. Making a rough translation of the caption on the photograph, I read the following:

> *Among the curiosities of the 1951 Mille Miglia was the participation of Nuccio Bertone, the owner of the Torino coachworks, who raced behind the wheel of a Ferrari barchetta built by his competitor Touring. In reality, Nuccio Bertone, as a sportsman, greatly appreciated this car and had already in the previous year's edition driven the same car to an honorable fourteenth place, third in the category Sport 2000. In this 1951 edition, however, he was forced to retire shortly after passing the Ravenna control.*

So Bertone had not only entered the 1950 Mille Miglia, but also had done rather well. Together with the factory's endorsement, this seemed to confirm that Bertone was the first owner for 1950 and 1951, and that Giletti must have followed him in 1952. But who was the Bruno Martignoni that my factory information referred to as the second owner? Was he the third? Was it him who had the top put on? Even as more questions were being raised, now I was getting somewhere. Obviously, 0046M had covered a lot of ground in the 16 years prior to my buying it and I wanted to know all of the details.

The European history seemed formidable and I felt I would have to rely on secondary sources and the work of others for that, so I concentrated on the US history. As I was trying to trace the Lotus 11 for my transmission, my calls were to people in Birmingham who had been involved in the sports car scene in the early '60s, so I also asked about a little red Ferrari with a Chevy engine and a big racing fuel cap in the middle of the rear window. When I found out about Maurice Blevins, I asked about him too. If I could find him, he could tell me who he bought it from and so on, but invariably I had no luck, usually being referred to someone else who might know. I spoke to such local notables of the day as C.K. Thompson, who had campaigned a D-Type Jaguar, and Tommy Charles, a popular Birmingham DJ who had raced a Chevy-powered Ferrari Mondial, but no one knew of my car or Blevins.

[18] p. 149. Zagato was, and still is, an Italian *carrozzeria*.

Homebound as I was, the telephone became my primary research tool, but since my right hand was in a cast, having the phone in my left hand meant there was no way to take notes. That I solved by getting an attachment from Radio Shack for my cassette recorder that allowed me to record my phone conversations, a practice I continued even after I regained the use of my right hand, and I have recordings of all the significant telephone interviews I conducted from then on. Writing this over 30 years later, I am able to listen to those conversations again with those who are since deceased and find valuable information that escaped me at the time.

Eventually, through the referrals I was told to contact Gerald Roush in Atlanta, a Ferrari historian and publisher of the *Ferrari Market Letter*. Word that I would be in touch had reached him and Gerald was not surprised when I called him one day in March 1988; in fact, he had done some preparation and reviewed his records. It turned out that he had seen my car around 1959 or 1960 when he was a teenager while mowing his parents' lawn in Birmingham. He had heard the sound of the engine over his electric mower and abandoned his chore to go see what it was when the car came to his neighborhood for a visit. According to Gerald, mine was the first Ferrari he ever saw, a seminal moment which set him on a lifetime avocation that became his livelihood, and he devoted a cover story to the tale in the February 4, 1989, issue of *Ferrari Market Letter*. Later, in 1962, he had seen an advertisement in the *Birmingham News* offering a Ferrari for sale and had gone to investigate. It was my car again, and he had taken several photographs, copies of which he sent to me. Perhaps it was the inspiration of 0046M that contributed to his buying a 166 Mille Miglia of his own as his first Ferrari: 0068M.

Gerald and I compared notes on what we each knew and discussed the confusion over who had been the first owner of 0046M, since Anselmi and Boscarelli had named Emilio Ciletti, and Stan Nowak had published an article where he had listed Bruno Martignoni as the first.[19] We both agreed that I held the trump card with my letter from Ferrari naming Bertone, which was straight from the horse's mouth, as it were.

As for the current owner, Gerald said the whereabouts of my car had been a mystery to those who tracked the old Ferraris, although there was speculation that a Birmingham man named George Wright had it hidden because he had coyly alluded to knowing where it was and knew details such as its now being a *berlinetta*, that a Chevy engine was in it, and that the Ferrari engine was "a basket case." More importantly, though, Gerald definitely knew who one of the past owners was, a man named Chuck Saffell[20] whom he had identified because he had recognized my car in slides that Saffell had sold to John Apen, along with other Ferrari materials. Apen, he said, was one of the owners of FAF Motorcars, the Ferrari dealer in Atlanta. Furthermore, according to Gerald, Saffell's photograph was in the *1957 Ferrari Yearbook*, so that was impressive.

My next call was to John Apen at his office at FAF. John listened as I explained who I was and what Gerald had told me about his connection with my Ferrari. While he was unaware of its serial number, he confirmed that he did have the slides of the car and that it had indeed belonged to Chuck Saffell, who was now in Florida. The slides were among many others he had acquired from Saffell in February 1983 and he offered to look for them and his notes, and then get back to me. A few days later we were back in touch and John had found 15 slides showing my car and had found Saffell's address. He was living in Palmetto, Florida, and owned at the time he had dealt with him a business named Competition Motors. Referring to his notes, John described the conversation he had with Saffell, speculated that he had imported the car into the US, and arranged with me to make prints from the slides and send them on.

A call to Information for Palmetto, Florida, gave me Competition Motors' telephone number where I learned Saffell had retired, but when I gave the password "Ferrari" I was given his number and was soon dialing him. This was an exciting moment for me since I was about to speak to the

[19] "The First 100 Cars: Part II," *Cavallino*, No. 2, p. 9.

[20] Pronounced SAFF-ul.

first former owner I had found – a little like finding a long-lost brother. Saffell – Chuck – answered and we bonded quickly. "How is the old girl?" he asked when I told him I had 0046M. Hard upon that came the question I would invariably hear as I tracked down my predecessors: "How did you find me?" "Not easy," became my usual reply. Our conversations and Chuck's part of the story will be covered in a later chapter, but pursuing the trail I asked him from whom he had bought the Ferrari and all he could tell me was that it was from a dealer in Detroit around 1957 or 58. Asked who he had sold it to, he said it was someone in Birmingham but he had trouble recalling the name, and when I suggested Maurice Blevins he agreed. Then Chuck contributed some rather astounding information:

> *Sold it to a kid in Birmingham who only had one leg. He was a cripple, oh he had one wooden leg.*

Now I had a new mystery, but I had a trail leading from Tuscaloosa to Detroit. Beyond there, I had no path, so having subscribed to Gerald's *Ferrari Market Letter* in the meantime, I went fishing with an advertisement, hoping to get a nibble. It ran under Miscellaneous Wanted in the April 2, 1988, issue, followed by my address and phone number:

> *166 MM FIVE-SPEED GEARBOX for restoration of 0046M. Any bits and pieces may be useful. Also need help with its history in Europe and also Detroit area around 1957. It is a Touring barchetta with a fastback roof added later by Zagato. CAR IS NOT FOR SALE!*

Right away, I received a letter from Stan Nowak – well known Ferrari historian, restorer and wheeler-dealer. Straight off, Stan contravened my admonition by saying he would like to buy the car if I were to sell it, but he went on, trying to be helpful by giving me what he knew about the early history. Unfortunately, he had nothing new to add, but he was a good contact in the Ferrari world into which I was now making my way. Stan and I corresponded and spoke a number of times before his untimely death in 1991.

Soon after Stan's letter, I got a phone call from Fred Leydorf who had seen my advertisement, and he really opened up new prospects. Fred had been an engine designer in Detroit for Ford for 10 years and with AMC for eight years prior to that. He was also one of the most technically versed people in the US on Ferrari engines, had published articles on his restoration work and was a prominent member of the Ferrari Club of America. What Fred told me was that, like Gerald Roush, his enthusiasm for Ferraris had stemmed from an encounter with my car. In 1956 he had gone to New York State to watch the Watkins Glen races and saw it run there, falling in love with it. He said he had a vivid memory of how the engine sounded at speed, with a high-pitched scream. His recollection was that it was entered by a Robert Publicker, and yes, he had a photograph he took that day!

Fred had another find for me too: a friend had photographed my car at Sebring in 1958 and when he saw those slides, he had recognized his old flame. Those images he still had, as well as his own from Watkins Glen, and he readily agreed to make copies for me. Indeed, he had been wondering where the Ferrari was all along and from the *Tipo 166* book he had deduced that it was the Zagato modified car, so when my advertisement appeared he was quite pleased to get in touch with me and wanted to hear all about it.

Fred's tip that my car had raced at Watkins Glen opened another door for me to investigate. Placing a call to the Watkins Glen office on April 25, I discovered that they had founded a racing research center there and was told to call Bill Green, who was their historian. Bill was quite receptive to my call and right away went to work to see what he could find for me from the Glen's 1956 records. The next day Bill called me back with good news. It was not Robert Publicker who had entered my car. Publicker had entered a Ferrari Testa Rossa that was a no-show, but it was John Sharpe from Buffalo, New York. He had much more for me, which I will reserve for a later chapter.

That night, after talking to Bill, I called Buffalo Information and got the number for John Sharpe. Minutes later I had him on the phone, recording as usual, and I introduced myself as the owner of his former Ferrari. He was surprised and delighted, and his wife in the background kept wanting to know how I found him. We had a long conversation, revealing many details and filling in blanks, chief among which was that the car had come from Luigi Chinetti, the US Ferrari dealer, and that Sharpe had acquired the car from him with the assistance of a fellow Buffalo resident, David Leopold, whom he thought was now living in Stuart, Florida. Unfortunately, he was hazy on where the car went.

I was feeling proud of myself for what I had accomplished in my research, so using what I had learned, I wrote an article in May 1988 titled, *A Short History of 0046M*, and submitted it to the Ferrari Club of America magazine, *Prancing Horse*. It was published in the 3rd Quarter 1988 issue, Number 88, page 32 (Appendix 13). By the time the article was published, however, it had become apparent to me I had been premature. There was much more to be known and I had made several errors. Even so, the article was accepted as gospel by many Ferrari historians and they perpetuated the misinformation for a long time, since I shared nothing more publicly from then until now.

On June 11 I returned to the trail, called Information at Stuart, Florida, and made a call to David Leopold. David turned out to be a fascinating character with a copious memory who had led an interesting and somewhat tragic life. He had such a wealth of information and I enjoyed talking to him so much that I called several times over the next few months, recording it all. It turned out that it was actually him who had bought the Ferrari from Luigi Chinetti, who had imported the car, then sold it to John Sharpe, and after re-acquiring it from Sharpe, he had sold it to the deceased Delavan ("Del") Lee in Detroit who had dealt in used exotics at his business, World Sports Car Center.

With the identification of Del Lee, I had found where Chuck Saffell had bought the Ferrari in Detroit and completed the ownership trail in the US. That is, except for finding the elusive Maurice Blevins with his wooden leg who seemed to have vanished. Indeed, I looked for Blevins off and on for the next 20 years without success.

Meanwhile, the advertisement I had placed in *Ferrari Market Letter* had reached farther and with greater success than I could have expected. After I sent off the *Prancing Horse* article, I received a letter from Antoine Prunet in Paris, a renowned author and highly regarded Ferrari historian. In part, he said,

> *Dear Mr. Westfaul,*
>
> *Although I had been tracking 0046/M's history since a fairly long time, I had no idea of its present whereabouts until I came across your ad in the Ferrari Market Letter. It is a good surprise.*
>
> *Here enclosed is the print out of my present knowledge on the early life in Europe of your interesting Ferrari 166 MM but no gearbox I regret.*

With one fell swoop I then had virtually the entire history of my car from its first owner to the present. Not only did Prunet have the list of European owners, he had also compiled the car's competition history over there, so this was a godsend. I saw now that I owned a really significant Ferrari. I could also see the progression of ownership in Italy definitively, going from Bertone to Giletti in 1952, to the previously unknown Luigi Bosisio in 1953, to Martignoni later in 1953, and then reappearing at Chinetti Motors in America. About the addition of the top, Prunet was less knowledgeable, but he speculated from the 1953 date in Boscarelli's book that it must have been done by either Bosisio or Martignoni.

To settle the question of who commissioned the top, I enlisted the aid of a translator, Mary Virginia Reese, a University of Alabama graduate student who spoke fluent Italian, and wrote to Zagato on June 29, 1988, asking that question and sending photographs. In time came the following reply:

> *Dear Mr. Westfaul,*
>
> *We have read your letter of June 29 with great pleasure and we rejoice with you over the splendid Zagato 166.*
>
> *Shortly the book "LE FERRARI DI ZAGATO" will come out. In this book, made in close collaboration with Zagato, you will be able to find the history of your car.*
>
> *Actually, we had sought out all the owners of the still*

duncan hamilton
Marston Magna
Nr. Yeovil
Somerset
BA22 8DH
U.K.
27 July 88
Modena Co., Ltd.
29/1 Soi 4 Thapae Road
Chiang Mai 50000 Thailand
JESS G. POURRET
MANAGING DIRECTOR
31 DEC 1990
WILLIAM GREEN
3211 Moreland-Beaver Dams Road
RD 1, Box 58
HISTORIAN
Watkins Glen International
Glen Region, SCCA
Watkins Glen Racing Museum
Deputy Archivist,
Sports Car Club of America
May 31, 1988
Milano,li 27 settembre 88
Egregi Signori
Darrell e Elisabeth Westfaul
Dott. Ing. CARLO FELICE BIANCHI ANDERLONI
FRED LEYDORF
AUTOMOBILE CLUB VARESE
raccomandata
prot.N. 1483
SEP 28 1991
21100 VARESE V.LE MILANO 25
TEL. 285150 TELEG. AUTOCLUB
EGREGIA SIGNORA
DARRELL WESTFAUL
P.O. BOX 032052
TUSCALOOSA,ALABAMA 35403
U.S.A.
RICHARD F. MERRITT
2220 N. Vermont
Royal Oak, Mich. 48073
club of america
Yours.
Antoine Prunet
Antoine Prunet
21, rue de Téhéran
75008 Paris
France

existing Ferrari Zagato cars, and we thought your car had been destroyed.

You will find your name included in the historic register of Zagato's Ferrari cars.

This was exciting news – the story of my car was going to be in a book! And at last my reach was now extending into Italy with Ms. Reese's help. Another letter dated September 27 arrived from Michele Marchianò, Zagato's official historian and the author of the book. He began:

Dear Mr. Westfaul,

The book Ferrari by Zagato *has recently been published in Italy. Your car, for which we compliment you, is amply illustrated in this book. This car, originally a Touring, enjoyed a glorious sports history, first with Enrico* [sic] *Giletti, and then, after its transformation into a coupe by Zagato, with Luigi Bosisio.*

... The information on your car in the book came from Luigi Bosisio, the Milanese driver from the scuderia Sant Ambroeus, who supervised the transformation by Zagato at the end of 1952.

This confirmed that Bosisio was responsible for the body modifications. Of course I had to have the book, so I found a source and ordered it right away. When it arrived, I was delighted to see a whole chapter devoted to 0046M and moreover to discover that it was one of only nine vintage Ferraris with Zagato coachwork, adding even more luster to its pedigree. To give the book – and my car – some publicity, I wrote a review and submitted it to *Cavallino* magazine, shamelessly pointing out Marchianò's coverage of a previously unheralded Zagato Ferrari, 0046M. My review was published in the April/May 1989 issue of *Cavallino*, Number 50, page 19.

As helpful as Marchianò's book was, I wanted more information, so with Ms. Reese as my interpreter and spokeswoman I again turned to the telephone. We found the number for Zagato, called them and then we were put in contact with Marchianò. He was a charming man who enjoyed flirting with Ms. Reese, but he had put all he knew into his book. Luckily, he gave us the telephone number for Bosisio, whom we then called, and had a lengthy, informative and recorded conversation with him, which will play a part in his forthcoming chapter. Bosisio had provided the photographs for the Zagato Ferrari book and Marchianò was kind enough to send me two prints of them autographed by Elio Zagato, son of the founder and now the owner (see **[141]** and **[144]** on pages 141 and 143). One of Marchianò's letters is included in Appendix 14.

Around this time, I learned that each automobile registration from my car's period included information on the sequence of previous sales and that the local branches of the *Automobile Club d'Italia* kept those records as the *Ufficio del Pubblico Registro Automobilistico* (Office for the Public Registration of Automobiles). Since I knew that Martignoni was the last owner in Italy and that he lived in Varese, a call from Ms. Reese to Varese Information gave us the phone number and from calling the *Automobile Club d'Italia* there we were told how to order a copy of 0046M's last registration. That document is reproduced in Appendix 5. Over time, I acquired copies of the other owners' registrations from different sources, all of which provided specific dates and other information after tedious deciphering of handwriting in Italian.

August was approaching, and I began preparing for my trip to California for the *barchetta* reunion. In the previous chapter, I detailed the importance of that trip for meeting Tohru Horinouchi and finalizing the purchase of his transmission, but there was another objective. I had learned that Nuccio Bertone would be at the Pebble Beach Concours as an honorary judge while our group was there and I wanted to meet him. For that anticipated meeting, I had two copies made of the photograph on page 55 of Marzotto's *La Ferrari alla Mille Miglia* book, showing Bertone and copilot Mirko in 0046M during the 1951 Mille Miglia. My hope was to present Bertone with one and get him to autograph the other for me.

The reunion was a wonderful experience, but a working one

too. Being around so many 166 MMs, I could study them up close and see and photograph details for the restoration of my car. I had also agreed to cover the event and write an article for *Prancing Horse*, which subsequently appeared in issue 93, the 4th quarter of 1989. An invaluable bonus was the guest of honor, Carlo Anderloni **[28]**, who had designed the *barchetta* body and whose Carrozzeria Touring had built and finished the 166 MM cars forty years earlier. Carlo was an amazing resource and for a week we reveled in his stories and most convivial company. He spoke fair English, but his wife Maria's English was quite good. They were a delightful couple, and in a stroke of luck after the reunion, my wife and I ran into them at Yosemite National Park doing some extra sightseeing as we were. We toured the park together and said our goodbyes for a second time when we parted.

The reunion was a week packed with one outstanding event after another and scheduled to culminate with the Pebble Beach Concours. The evening before, we were hosted for dinner by Gil Nichol aboard his yacht in Monterey Bay and there I met Antoine Prunet, whose help had been so invaluable. Prunet had added to his research with more racing results and the details he learned from my *Prancing Horse* article. This he shared with me, having brought a printout for that purpose, and I was most grateful.

Our group assembled for the farewell luncheon on August 20 at the Pebble Beach Clubhouse during the Concours. I had my camera and was clutching a manila envelope containing the two photographs, but I had no idea how I was going to find Bertone among the throng of thousands. While we were seated and eating lunch, Sherry Gilbertson, Ed's wife, pointed him out to me as he passed our doorway in the hall, but by the time I had excused myself and

[28] Carlo Anderloni sits in Tohru Horinouchi's 166 MM. *Author*

[29] Carlo Anderloni (left) autographs the photograph of 0046M while Nuccio Bertone (right) looks on and waits to sign his name. *Author*

went out to look for him, he was nowhere to be seen. So, when we disbanded, I went back out to watch the award presentations and joined the masses pressed along the rope separating the commoners from the gentility as they drove up to receive their awards.

I knew Bertone would be on the judges' stand and so would Carlo Anderloni who was also an honorary judge. As I worked my way to the front, I could see them together on the left end but not under the awning among the numerous other judges and award presenters. This, I reasoned, was my best chance because there was no certainty of finding him afterward. Summoning my nerve, I stepped over the rope and walked up to one of the sheriff's deputies positioned to keep the likes of me in place. I held the manila envelope up as I approached and said officiously, "I have to deliver this to Signor Bertone," and he waved me on by! When I reached them, Carlo got Bertone's attention for me and I pulled the photographs out, showing them to him and gesturing from the photographs to him and back. Carlo took over at that point as translator and Bertone beamed as I explained through Carlo that this was a picture of him racing and that I now owned his car. It actually created a mini media event as cameras and microphones were thrust among us while this unfolded. Fulfilling my hopes, Bertone autographed one of the photographs for me and Carlo added his as well, a historic piece of memorabilia by two men associated with the genesis of my car. In **[29]**, Carlo prepares to sign while Bertone looks on at the right, green pen in hand,[21] and in **[30]** Bertone is speaking to an Italian journalist while I am in the background between them. The autographed image appears later in **[54]** on page 74.

Since he had to return to his judging duties, there was no time to talk to Bertone in depth as I would have liked. Indeed, I was never able to question him about his time with 0046M as I did others and had no response to my letters, but it was with him that my Ferrari's history began.

[21] Like Enzo Ferrari who had a trademark ink color (purple), Bertone's ink of choice was green. On the left, assisting with holding the photograph is Dr. Gian Beppe Panicco, press and public relations official for Carrozzeria Bertone. Dr. Panicco took my address and later sent me a copy of *Nuccio Bertone Portfolio* personalized with Bertone's autograph, again in green.

[30] Nuccio Bertone speaks to an Italian journalist while I stand in the background. *Author*

Chapter 4

1950-1951: THE BERTONE YEARS

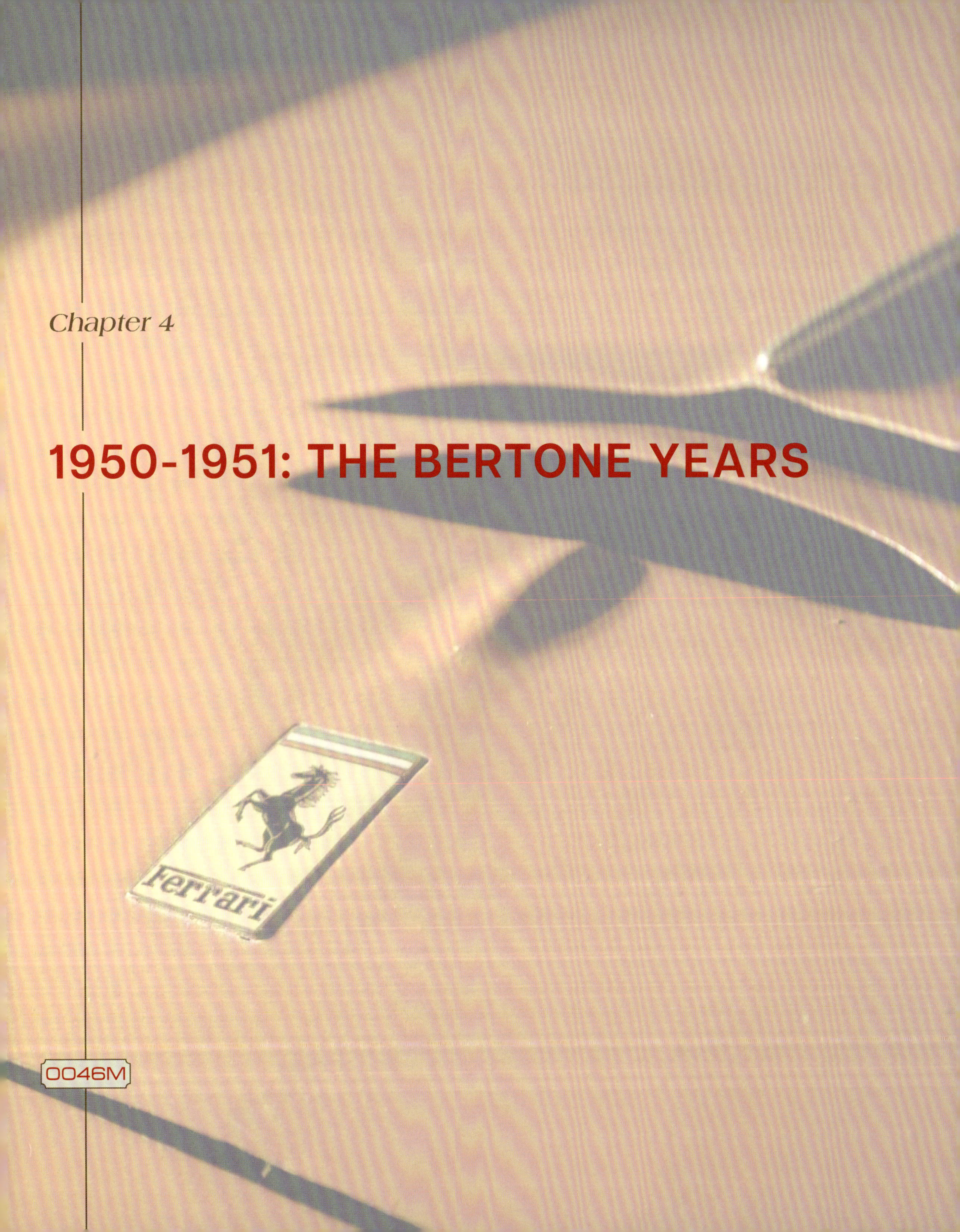

0046M

Giuseppe "Nuccio" Bertone was born into the culture of the automobile. His father Giovanni had founded his *carrozzeria* – or bodyworks – in 1912 and Nuccio arrived two years later. The business produced custom, hand-built bodies on the chassis of manufacturers such as Fiat and Lancia, growing in esteem and financial success until World War II interrupted. After quitting school short of having his degree in accounting, Nuccio Bertone joined the family business in 1933 at the age of 19. He learned from the ground up and soon designed and built a car on a Fiat chassis. His greater value to the company, though, seems to have been as a sales representative, traveling all over Italy to visit auto dealers and manufacturers, garnering orders and collecting payments for Bertone designs. Through that experience, he made valuable connections in the Italian automobile industry and honed business skills that would serve him well in the future.[22]

During the war, all automobile factories were pressed into service producing war material, which consequently resulted in their being bombed, so after the war all the Italian auto manufacturers were damaged to some degree. For a period, the Bertone works survived by rebuilding old or damaged vehicles. Eventually Gianni Agnelli, whose family owned Fiat, convinced the allied occupation forces to prioritize the reconstruction of his factories. Soon, there was again a supply of chassis and engines for the coachbuilders such as Bertone, and there was a ready market for autos, both for the masses and for the well-to-do. From 1950 to 1960, as Dario Moretti observes, "the automobile in Italy took on a remarkable mythical stature, in large part due to the establishment of the model of the fast, luxurious, state-of-the-art automobile."[23] Thus Bertone and other *carrozzeria* such as Touring found a foothold, along with niche manufacturers such as Ferrari.

Auto racing competition had long been popular in Italy and Nuccio Bertone had avidly followed it before the war. When the chaos of reconstruction had settled, racing returned in an explosion of events all over Italy and Sicily with a menagerie

[22] Luciano Greggio, *Bertone*, pp. 72-75.
[23] *Gilco: Le macchine di Gilberto Colombo*, p. 3.

[31] Father and son, Giovanni and Nuccio Bertone, a 1960s photograph. *Wikimedia Commons*

[32] Carrozzeria Bertone in the mid 1950s. *Wikimedia Commons*

of entrants from the family Fiat, to modified variants, to purpose-built race cars. Because there were few dedicated race tracks, multitudes of Italians lined appropriated city streets and rural roads to watch prewar heroes like Tazio Nuvolari and Giuseppe "Nino" Farina compete with a fresh crop of gentleman racers who were often the next generation heirs of industrialists. Nuccio Bertone was one of those.

At the age of 32, in 1946, Bertone began his racing career at the Asti circuit in a Fiat 500 with a Bertone-built roadster body adequate for the purpose. It was too heavy to achieve any success, but as Rob de la Rive Box and Richard Crump have pointed out, his entry into racing gave him a perspective on the competition side of the auto industry and renewed important pre-war contacts, but at the same time his participation belied to those contacts the dire postwar financial state of the Bertone coachworks.[24]

In 1947, Bertone allied with Vittorio Stanguellini and began racing cars built by him. These were not only lighter but the Fiat 750 engine in his car was modified to produce more power, and with a competitive machine he began producing respectable results, recording a 1st and a 2nd in class. In 1948, he moved up to a Stanguellini with a modified Fiat 1100 engine and achieved considerable success with two 1sts overall, three 1sts in class, two 2nds in class, and one 3rd in class. According to *Stanguellini: il "mago" dei motori*, Bertone "was a close friend of Stanguellini and... won his first races driving a 1100 Sport owned by Stanguellini."[25]

It seems Bertone managed to go racing without the expense of buying and maintaining a car, something he needed to do because he was keeping his racing a secret from his disapproving and financially ailing family. Here his salesmanship may have come into play because it appears to have been a reciprocal relationship with Stanguellini. As much as Bertone needed a car he could win with – discreetly – Stanguellini needed to sell his cars and Bertone could offer help with that. Again from *Stanguellini*, "1947 is the year of the first road car of Stanguellini: a *berlinetta* 1100, bodywork done by Bertone, with a closed coupé body. This elegant vehicle, even if it was made for an audience that loved mostly luxury, would be used for racing on several occasions and with good results."[26] And from *Bertone*, "A few dozen Stanguellini *berlinettas* were produced and they were a frequent sight in the amateur competitions of those years; off the track, they were often spotted in the smart part of town and in a variety of exclusive watering holes where they stood

[24] *The Automotive Art of Bertone*, p. 9.
[25] Dante Candini and Nunzia Manicardi, p. 50.

[26] Ibid., p. 50.

[33] The Ferrari factory 1947 with a 125 S in the entrance. *Ferrari S.p.A.*

DALTON WATSON FINE BOOKS

NEWEST RELEASES

Dalton Watson Fine Books
Glyn and Jean Morris
www.daltonwatson.com
+1 847 274 5874
info@daltonwatson.com

WAYNE CARINI: STEERING THROUGH LIFE

Wayne Carini

176 pages • 350 illustrations $59

Wayne Carini is the world's favorite classic car enthusiast and in *Steering through Life*, this beloved and respected gearhead looks back at a lifetime pursuing his automotive passions. From learning about cars at his father's side as a child through a career hunting the rarest and most desirable automobiles ever built, Wayne writes about his most cherished motoring memories. *Steering through Life* puts you in the passenger seat with Wayne at the wheel talking about his favorite cars, most exciting adventures and revealing some tricks of the trade. How did the internationally acclaimed TV program *Chasing Classic Cars* get started and what are the cars that he regrets letting get away? This book answers all those questions and more. With hundreds of full-color images, entertaining stories and countless insights, this is the perfect book for car lovers of all ages.

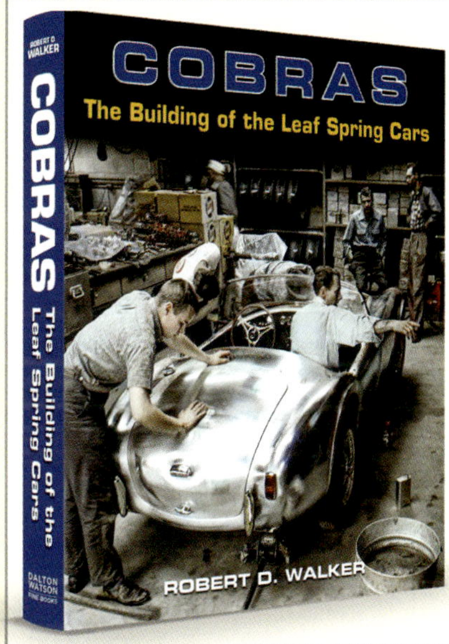

COBRAS: THE BUILDING OF THE LEAF SPRING CARS

Robert D. Walker

400 pages • 784 illustrations $175

COBRAS, The Building of the Leaf Spring Cars is a road map to how the 1962-1964 leaf spring Cobras originated and were put together by AC Cars in England. The book also details how the cars were completed in the United States by both European Cars in Pittsburgh, Pennsylvania and especially by Shelby American in Venice, California. The thirteen chapters cover all major Cobra components consisting of chassis construction, aluminum body forming, engine development, driveline components, upholstery, and finishing techniques. Each chapter is illustrated with dozens of rare original black-and-white factory photographs and background material, such as component illustrations and parts numbers. The research information that is included in these pages is sourced from well over a dozen of the most knowledgeable Cobra researchers, historians and restorers in both the United States and the United Kingdom.

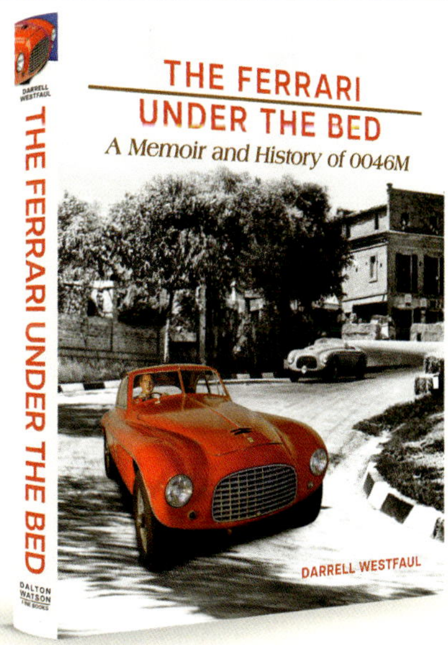

THE FERRARI UNDER THE BED

Darrell Westfaul

288 pages • 328 illustrations $125

Darrell Westfaul owned Ferrari 166 Mille Miglia 0046M for forty years, buying it as his first car when he was twenty-one. What followed were over five decades of searching for its past owners and its history with them. This book is Westfaul's gathering of obscure and previously unknown information about a Ferrari that had led a life of successful competition and then traveled widely in retirement, reaching three continents and crisscrossing the Atlantic. Had it simply remained as originally built, it would have been a rare and important car, but the changes it underwent have marked it as unique and historic. This book delves into those changes and the reasons behind them, accompanied by scores of previously unpublished photographs. Not only is there a detailed competition history, but also the words of men who raced it, transcribed from their own voices. In addition, there are conversations with other men and women whose lives intersected with the Ferrari. But, in a larger sense, this book is the story of a man and his car and their time together.

out from the herd thanks to their sleek and unquestionably elegant bodywork."[27] So there is evidence of a quid pro quo – Bertone producing a marketable car in exchange for the use of a racer and being successful with both.

The year 1949 began well for Bertone with a 3rd in class for his 1100 Sport Stanguellini at the Targa Florio, but other than another 3rd in class at the Coppa Inter-Europa, the season produced nothing of importance. The nimble 1100 S cars were still able to garner class wins in other hands, but there was only one 1st overall. The rising star was the 2-liter 166 Mille Miglia Ferrari, and Bertone must have cast an envious eye at its prowess.

Enzo Ferrari had built his first two cars in 1940 as Auto Avio Costruzioni, since he was restricted by contract with Alfa Romeo, his former employer. These were built by joining two Fiat engines end to end while the mechanicals were clothed in a body by Carrozzeria Touring. The advent of the war brought development to a halt and Enzo turned to the manufacture of machine tools, which would come in handy once the wreckage of the war was repaired. Able to build an entire engine and drivetrain in house and now free to use his name, Enzo formed Auto Costruzioni Ferrari

[27] Greggio, p. 118.

in 1945. The first Ferrari *automobili* came in 1947: two V12-engined sports cars, the 1.5-liter Tipo (type) 125 S, while his competition made do with 4, 6 or 8 cylinders. One of them sits in the entrance to the Ferrari factory at Maranello that year in **[33]**. The number 125 indicated the displacement in cubic centimeters of a single cylinder, now a long-standing Ferrari system.

Originally designed by Gioachino Colombo as a supercharged engine for formula racing, the new engine was short on performance when normally aspirated for sports car use, so the cylinders were soon bored to a capacity of 1.9 liters and two more cars became the 159 S. More engine modifications and enlargement to a fraction under 2 liters created the cycle-fendered 166 Spyder Corsa (SC) for 1948.[28] The 166

[28] A *spyder* is a roadster or open two-seater car. *Corsa* means racing.

[RIGHT: 35] 0002M, the first 166 Mille Miglia, at the 1948 Torino auto show. *Revs Institute*

[BELOW: 34] Seldom photographed behind the wheel of his racing cars, Enzo Ferrari appears pleased *Ferrari S.p.A.*

engine provided more power and more success through the 1948 racing season, including an outright win at the highly regarded Mille Miglia race, which brought much publicity and prestige. Of 77 race entries in 1948, there were seven 1st overall wins, eleven 2nd places, five 3rd places for Ferrari 166s, and they placed a total of 39 times.[29]

On September 15, 1948, after building not much more than a baker's dozen racing cars, Ferrari exhibited a new model at the Turin XXX Salone Internazionale dell'Automobile, the 166 Mille Miglia, or 166 MM, named in honor of the race victory earlier in the year by Clemente Biondetti **[35]**. The engine was the latest 2-liter V12 and there were chassis

[29] Tito Anselmi and Lorenzo Boscarelli, *Ferrari Tipo 166: The Original Sports Ferrari*, p. 130.

improvements, but what attracted attention was its roadster body, newly designed by my late friend Carlo Anderloni and built by his Carrozzeria Touring. His mission, Anderloni said, was "to succeed in personalizing the looks of the Ferrari, which until then had been rather anonymous."[30] And so its shape was a radical departure from common practice, even controversial. To further quote Anderloni:

> *Mr. Giovanni Canestrini, one of the best journalists and the engineer who wrote for the* Gazzetta dello Sport, *described the car as "disconcerting" because there was no way of comparing it with previous models. He said it didn't look like a Spyder but more like a small boat. And that's how the car got its name "Barchetta." This sports car has a revolutionary body design which is not at all traditional, without the low clearance underneath, and the wide lower part becoming narrower as it tapers upward. With the "Barchetta," the widest part is at the horizontal midline, and its lower portion tapers to resemble the hull of a rowboat. It also appears to sit high off the ground.*[31]

[30] Carlo Felice Bianchi Anderloni, *"Barchetta... a Myth," Registro Internationale Touring Superleggera*, April 1998.

Undoubtedly, Nuccio Bertone was at the Turin auto show and admired the new Ferrari, but he must have admired it even more once it began showing its racing potential. While Bertone was having a lackluster season with the Stanguellini in 1949, the new 166 MM was winning races. The first major win was another 1st place by Clemente Biondetti in the Mille Miglia and then one by Luigi Chinetti at the 24 Hours of Le Mans. Indeed, by the end of the season, Ferrari 166s had won 1st overall 20 times: 9 on racing circuits, 7 in hillclimbs and 4 in road races – one of the best records ever for Ferrari sports cars. It can honestly be said that the 166 Mille Miglia was the foundation on which the Ferrari legend was built.

Not wanting to miss the "little boat," I suspect Bertone approached Enzo Ferrari and dangled the prospect of designing and producing a marketable car for the affluent, as he had done for Stanguellini, and negotiated for the use of a 166 MM in the meantime. However, including the 1948 Turin show car, Ferrari had built only ten 166 MMs through 1949, and while the success of the model had brought in many orders for the *barchetta* racing cars and some *berlinetta* versions, by the end of April 1950 only eight more had been completed.[32] Hence, asking Enzo to forego the revenue from one unit would have been asking a lot from a nascent factory. But as fiscally conservative as Enzo was, he was practical and concerned only with the end result.

[31] C.F. Bianchi Anderloni and Valerio Moretti, *Ferrari 166 Mille Miglia Barchetta*, p. 12.

[32] By my count from the Registry on *barchetta.cc*. The 166/195/212 Inter touring cars accounted for most of Ferrari's 1950 production. Supposedly, the name "Inter" honored Scuderia Inter that raced the earliest Ferraris. Inters were generally less powerful coupes with odd serial numbers having digits between (inter?) the even serial numbers of the racing cars.

While the past few years had been a phenomenal success for Ferrari, there was strong competition from other manufacturers, such as Lancia and Alfa Romeo, and Enzo realized he was at a growing financial disadvantage. He understood their business model, which was to build an image though racing and use that to sell profitable touring cars, which in turn provided the funds to support racing, leading to more success, prestige and sales. Most of the events at the time were about endurance – racing long distances over ordinary roads – and winning meant the manufacturer produced efficient, durable, and therefore desirable, cars.

While Ferrari had been offering the odd-numbered Inter series of touring cars, they were essentially the racing chassis clothed in a variety of bodywork and not a distinctive line. Besides, they were a distraction from the real purpose of building pure racing cars. Later on, Enzo put his dilemma succinctly:

> *The productive development of my business can be of interest to me only if it is run by others. One solitary thing drives me – to be able to dedicate myself totally to prototypes and to racing cars, but only under conditions of absolute independence...* [33]

So, it would have been an opportune time for Bertone to promote his company to Enzo as the solution to his need for "others" to handle his touring car business while he built the brand through racing.

Certainly, Ferrari did not approach Bertone, since one came to Ferrari and not he to them; plus, Enzo was known for his hard dealing and had the tendency to play one off against another, as he did with drivers. While Touring was currently the principal supplier of Ferrari's coachwork, there were other *carrozzeria* who had or would build their interpretations on Ferrari chassis, such as Allemano, Stabilimente Farina, Ghia and Vignale. "Pinin" Farina[34] said later, "It was obvious that he had been looking for a bodymaker 'for himself,' with whom to establish a new sort of harmony."[35] The Bertone works had a pedigree as good as any, and Ferrari surely knew of the pre-war success of Giovanni Bertone. Being keenly aware of the racing scene, he would also have known that the son, although an amateur racer, had achieved a measure of success in competition, something he needed for publicity to sell cars. Surely, if Nuccio came to him with a proposal that would benefit his business, Enzo would have listened – and Bertone had learned salesmanship.

Factoring into Bertone's proposal was likely the Ferrari Inter chassis numbered 0057S,[36] completed on August 1, 1950. Clearly, Carrozzeria Bertone acquired the chassis to build on it a body of its own design, but the seductive vision presented to Enzo could have been that of eager buyers clamoring for copies and consequently the sale of many chassis for Ferrari. This could have been the pitch Bertone made to Enzo for the loan of a 166 MM. Apparently, Ferrari was not able to fill the order for an Inter chassis until it was too late for the spring auto shows in Geneva (March) and Turin (May), but Enzo might have been persuaded by strength of reputation to grant the loan of a *barchetta* in advance for the 1950 racing season.

Whatever the process, Bertone obtained the use of 0046M from Ferrari and it carried the license plate PROVA TO 99, indicating a test (*prova*) car from the Province of Torino where he lived. The PROVA plate was only issued to constructors or manufacturers as a way to operate cars that had not yet been sold, and while no records were kept of those plates, Carrozzeria Bertone was certainly eligible for them. Typically, the Ferrari factory's PROVA plates were MO for the Province of Modena, or BO for the Province of Bologna,[37] so it would seem that the car was in Bertone's possession but not his ownership.

0046M was 18th of the series of 25 166 Mille Miglia *barchettas* by Touring, the even serial number denoting a competition car. One of the earliest factory sales brochures for the

[33] Stan Grayson, ed., *Ferrari the Man, the Machines*, p. 32.

[34] This was Battista Farina, nicknamed "Pinin," who legally changed his name in 1961 to Pininfarina. His older brother Giovanni established Stabilimente Farina before him, which was absorbed by Battista's Carrozzeria Pininfarina.

[35] *Ferrarissima*, vol. 1, p. 63.

[36] Also seen as 057S, but both the chassis and the engine are stamped 0057S.

[37] *Cavallino*, No. 66, p. 6 and No. 73, p. 14.

model is reproduced in Appendix 1. To achieve rigidity and light weight, Touring used their *superleggera* (super light) technique of welding a superstructure maze of small steel tubing above the chassis platform of two large, cross-braced main tubes[38] and then welding together over them a patchwork of soft aluminum sheets that had been hand formed by hammer to match a wooden "buck," or full-size model. The aluminum skin was then wrapped around the tubular framework at the edges to unify the body and chassis. The rough surface was then coated with filler, said to be a mixture of talc and resin,[39] hardened, and then sanded smooth.

Like most racing Ferraris since, they were all right-hand drive.[40] As roadsters, they had no provision for a top and usually had two small, low windscreens. A few were *lusso* or luxury models, as was the Turin show car, and Anderloni described their differences as "limited in the majority of instances to using English leather instead of imitation product, to more powerful headlights (Marchal), to the typical [leather] trimming around the cockpit and from a color different from red... "[41] 0046M, on the other hand, had the austere cockpit of the *corsa* cars – a machine meant to race. The bucket-style seats were most likely covered in vipla, an imitation leatherette material, probably brown; there were no carpets; and of course the body was red, *rosso corsa*, the blood red Ferrari racing color of the early years.

I acquired the factory Assembly Sheets for 0046M through Ferrari of North America in 1991 and they gave the specifications of the car in its stages of completion and subsequent modifications, along with dates; these can be found in Appendix 2. This was a busy time for Ferrari since the all-important Mille Miglia race was coming up on April 23rd and there were many cars to prepare.

[38] Sourced from the Gilco company in Milan. Ferrari only manufactured the drivetrain.

[39] Michael Dregni, *Inside Ferrari*, p. 70.

[40] Sitting on the right and shifting with the left hand leaves the (usually) dominant right hand free to control the car. Also, with clockwise direction of travel being the norm on European racing circuits, the pits are usually located on the driver's right, and having him on the car's right side is faster for assistance or a driver change.

[41] Anderloni, *"Barchetta... a Myth."*

The assembly of the engine was completed on March 16, 1950, with a single Weber 36 DCF carburetor[42] and a compression ratio of 8.5:1, and then it was run on a test stand on March 18. As a rule, Ferrari redacted performance data from the Assembly Sheets when they were made public, so no results can be seen, but the early 166 MM sales brochure reproduced in Appendix 1 lists 130 horsepower at 7,000 RPM, while the information I received from Ferrari in 1966 claimed 136 horsepower and another 166 MM brochure has 140 horsepower,[43] but improvements were ongoing. The transmission – serial number 8 – was finished on March 29 and the output was geared for the ratio 9x42. 0046M's Certificate of Origin was dated April 6, 1950, so the engine and transmission were likely mated to the chassis by then along with the differential – serial number 0018 – which was geared differently at 8x40 **[36]**.

Anderloni describes the next step: "The routine procedure was simple enough: Ferrari ordered the bodywork here [Carrozzeria Touring in Milan] and dispatched the chassis to

[42] *Doppio Corpo Ferrari* – a twin-throat carburetor, each 36 mm wide, made by Fabbrica Italiana Carburatori Weber of Bologna for Ferrari.

[43] Richard F. Merritt, *Ferrari: Brochures and Sales Literature – A Source Book 1946-1967*, p. 41.

[36] The serial number and gear ratio stamped on the differential housing. *Author*

us. While the work was being carried out, he sold the car."[44] Except in the case of 0046M when it would be loaned.

I have no information on the date Bertone actually received 0046M, since the testing and delivery date to a client were left blank because, in truth, there was no client. One would expect he had it in time to familiarize himself with it before racing, though, which may explain why the car was back at Ferrari on April 21.[45] The Assembly Sheets have a notation on that date which may indicate the drum brakes received a final adjustment then after break-in, with a view toward the 1,000 miles (1,609 km) of racing ahead on April 23 – the first race of 0046M's career.

[44] Anselmi and Boscarelli, p. 56. Photos show a 166 MM *berlinetta* body being constructed over a driveable chassis.

[45] This date appears to apply to the completion of the differential but that was needed earlier to have a rolling chassis for the bodywork.

For the XVII Mille Miglia, Bertone chose as his co-driver and navigator Luigi Casalegno, whose racing career seems to have consisted of being his assistant. As Bertone said, "In the Mille Miglia I took co-drivers just as they came... students even. It didn't matter because in any case I drove the whole way, from start to finish."[46]

Photographs **[37]** and **[38]** were taken by Alberto Sorlini and given to me by Ferrari historian Jess Pourret from his archives in 1990 after a year of playing mail tag as he moved between his residences in Paris and Thailand.[47] In them, the two can be seen preparing to be released and start their race in the rain from Brescia, with Bertone behind the wheel on the right side in his peculiar helmet. As with many of

[46] Greggio, p. 127.

[47] The rights to the Sorlini photographs are now owned by Giorgio Nada Editore Archive.

the 166 MMs with twin windscreens, a piece of Plexiglass is used to close the gap between them to reduce rain and wind in the open cockpit. To further assist, small Plexiglass windwings are attached to the uprights of the windscreens on each side, an uncommon addition.

More features that help distinguish 0046M from other *barchettas* are the thin chrome rims around the headlights and the location of the plugged hole for hand crank access at the bottom of the front grille. The chassis of these cars rigidly and uniformly located the position of the engine and radiator so that a supplementary hand crank could be inserted through an opening at the bottom of the radiator and mate with a coupling on the engine, but when the hand beaten aluminum panels were welded together, the result was that no two bodies were precisely alike; hence, there were variations in the position of the bodies on the chassis that necessitated slightly different vertical locations for the access hole in the grille, and this can be seen varying by as much as a couple of inches from car to car. It is an overlooked way of visually identifying one 166 MM from another.

[ABOVE & OPPOSITE: 37-38] XVII Mille Miglia, 1950 – Bertone at the wheel with Casalegno in 0046M for the rainy start. *Giorgio Nada Editore Archive*

The car wears the race number 652, which indicates its scheduled start time, 6:52 AM. Cars were sent off at one minute intervals, the slowest leaving first to allow the pack to telescope together toward the end. It also meant that the faster cars, leaving last, had to pass literally hundreds of other cars, often on winding roads of dubious condition.

The photographs also plainly reveal the weather conditions that morning. As *The Autocar* reported in the April 28, 1950, issue:

> *It was one of the most difficult races ever held. Torrential rain and hail, mist and slippery roads, darkness and heavy traffic, wore down the nerves and the stamina of drivers and there were many accidents, some fatal* [3]... *Weather forecasts before the start were gloomy, but drivers cheered up when Saturday night broke clear with a bright moon and stars. The touring and small cars had a clear run over the first section during the small hours, but mist arrived later and as dawn broke and the fastest cars approached the starting line, a deluge descended and turned the first 150 miles* [241 km] *of road into a skating rink which provoked a series of disastrous crashes... One driver said that the countryside was "littered with Ferraris."*[48]

These were the days of drum brakes, which became dangerously unreliable when wet. Even so, images **[39]** and **[40]** capture Bertone and Casalegno at speed in Florence, after which they are stopped there at the Piazza della Libertà checkpoint **[41]**. In the open cockpit of 0046M, it must have been a miserable race for them, but in **[42]** they have found a dry stretch of road. Nevertheless, they persevered and at Ravenna (303 km) they were running 2nd in the 2000 Sport class. Through Pescara (604 km) and Rome (901 km) they lost ground and fell to 5th in class, but by Livorno (1,220 km) they had climbed to 3rd in class and finished the race in that position, also earning a respectable 14th overall with a time of 15:22:46.4. Indeed, through Florence (1,345 km) and Bologna (1,452 km) they had been running in 11th place overall.[49]

[39-40] XVII Mille Miglia – At speed through Florence.
© Archive Photo Locchi Florence

To put this in perspective, there were 375 cars entered and 162 failed to finish. There were 16 cars in Bertone's class, which included nine other 166 MMs and three of the 166 S cars. There were four other Ferraris in a larger engine class contending for overall results, one of which was the winner, one 2nd and one 9th. Another 166 MM finished 4th, making Bertone's the fifth highest finishing Ferrari. So, it was an auspicious beginning for the career of 0046M.

Although the Mille Miglia was a 1,000-mile (1,609 km) race, it began and ended in Brescia, 225 km away from Bertone's home. Racing as an independent without the support of a *scuderia*, or racing organization as were being formed,

---

[48] Andrea Curami, ed., *Mille Miglia Race: The Postwar Years*, pp. 27-28.
[49] *Auto Italiana*, April 30, 1950, pp. 94-114.

[41] XVII Mille Miglia – Slowing for the Piazza della Libertà control point in Florence. *© Archive Photo Locchi Florence*

[42] XVII Mille Miglia – Along an autostrada in the countryside. *The Klemantaski Collection, Corrado Millanta*

Bertone had typically competed in north Italy where he could "load the car onto the trailer and drive back to Turin overnight in order to show up for work on Monday morning, as if nothing had happened."[50] Even after his family realized what he was doing, he seldom ventured long distances to a race site.

A little over a month after the Mille Miglia, on June 4, Bertone entered 0046M in the II Coppa della Toscana, beginning and ending in Livorno, 420 km from Turin, and again his co-driver was Casalegno. It was a race of 680 km over public roads through six checkpoint towns, and for this race there were 181 cars entered. As in the Mille Miglia, there was a mix of classes with slow and fast cars, so there was a staggered start and Bertone's race number of 1211 indicated his afternoon starting time. Drivers were issued a printed route guide that described road conditions and features at measured kilometer intervals, the cover and first page of which is shown on page 65 **[43]**. About the first segment, it says: "Flat route of little difficulty, very fast. Excellent road."

In the program for the next year's III Coppa della Toscana there are several photographs from this race, one of which shows a woman wearing sunglasses assisting Bertone with the chin strap of his helmet. If not a girlfriend, she might be his older sister Maddalena ("Nene") and she is standing on the front row intently watching him start **[44]**.

From the start, Bertone and Casalegno led their 2000 Sport class through the checkpoints of Massa, Florence and Sienna, steadily adding about 30 seconds to their lead at each. Then as fate would have it, on the approach to Viterbo there was a railroad crossing where they were forced to wait for the passage of a train, allowing the Maserati of Sonio Coletti and Fabrizio Serena to catch them. At Viterbo they arrived with equal times of 03:42:00 each, and from there on Coletti and Serena took the lead, finishing in 05:54:46 to Bertone and Casalegno's time of 05:58:02, leaving them 2nd in class and 5th overall. In **[45]**, they have crossed the finish line back in Livorno, grimy but animated **[46]**. *Auto Italiana* called the win of Coletti and Serena "a beautiful

[50] Greggio, p. 117.

[44] Il Coppa della Toscana – Bertone and Casalegno on the starting line in Livorno. © *Archive Photo Locchi Florence*

1950
2ª COPPA
DELLA
TOSCANA
KM. 680
FILIPPI
ALQUATI
IPPOCAMPO
BOLDRINI
SCHWELM
COLETTI
SCOTTI
SERAFINI

LIVORNO - MASSA

Km. 68

Percorso pianeggiante di scarse difficoltà, molto veloce. Fondo ottimo. Si segnala:

Chilometri progressivi	Riferimento segnali Km.ci	
	S. S. N. 1 Aurelia, numeraz. cresc.	
1	—	**Livorno** - Viale Italia - **Partenza.**
5	316	**Livorno** centro - incrocio Aurelia - Viale Carducci - Cunetta.
6	dal 317	Fondo stradale ondulato.
8	al 319	
10	321	**Stagno** - Curva a destra - passaggio a livello.
11	322	**Stagno** ponti in rifacimento - rallentare - PERICOLO.
23	334	**Pisa** - 100 m. prima imbocco ponte sul-l'Arno cunetta.
44	355	**Viareggio** (ingresso) - passaggio livello e curva a sinistra.
45	356	**Viareggio** - curva stretta a destra - Attenzione.
46	357	**Viareggio** (uscita) - passaggio a livello.
49	360	Curva a squadra a sinistra - il percorso abbandona l'Aurelia e s'immette nella Comunale per il **Lido di Camaiore.**
50	—	Dalla Comunale sul lungomare del **Lido di Camaiore** - curva a destra - Attenzione.
56	—	**Forte dei Marmi** - Inizio lancio per tratto velocità - **attenzione segnali riferimento.**
57	—	*TRATTO VELOCITÀ.*
58	—	Fine tratto velocità, subito dopo viale fiancheggiato da siepi basse; attenzione a numerose cunette in corrispondenza di strade laterali.
65	—	**Marina di Massa** - curva a squadra a destra.
68	—	**Massa** (entrata) - Controllo e rifornimento lungo viale alberato centrale.

[43] II Coppa della Toscana, 1950 – route guide book, cover and first page. *Automobil Clubs Della Toscana*

tactical victory theirs, but on the other hand a sure chance to win slipped quietly out of hand for the great Torinese coachbuilder," and it commented, perhaps with irony, that he was "without suitable alerts in the last part" of the race.[51] Even so, it was another good result for 0046M on its second outing.

[51] *Auto Italiana*, June 15, 1950, pp. 57-58.

In the following month, the Assembly Sheets indicate that 0046M went back to Ferrari and by July 4 the ratio of the differential was changed from 8x40 to the higher 9x42 option. Perhaps this was to correct the mismatch between the transmission's odometer output of 9x42 and the differential, giving more accurate distance information, but increased performance was most likely the purpose. With the standard 8x40 axle ratio, the top speed of a 166 MM was 205 kph

[45] Il Coppa della Toscana – race finish at Livorno. *© Archive Photo Locchi Florence*

[46] Il Coppa della Toscana – after-race chatter. *© Archive Photo Locchi Florence*

(127 mph), while with the 9x42 ratio it was 220 kph (136 mph).[52] Bertone had been unable to keep up with the Maserati of Coletti and Serena, so he likely wanted more speed.

Other work completed by Ferrari then was on the electrical system: the horns and the starter motor were repaired or replaced. Adjustments were also made to the handbrake and the steering column, with the length of the latter fixed at 1,405 mm, better perhaps to suit Bertone's small stature. The accelerator pedal was adjusted, too, again possibly to suit Bertone. The heights of the clutch and brake pedals were mechanically adjustable, while the accelerator pedal height was changed by attaching a wooden block to its surface, visible in **[17]** on page 28, and seen on other 166 MMs.

On July 16, Bertone entered 0046M as car 147 in the IV Coppa d'Oro delle Dolomiti, a road race of 304 km through the mountainous territory of the Dolomites in northeast Italy on winding, treacherous roads that were often unpaved gravel. Figure **[47]** shows Bertone and probably Casalegno in one of the urban passages while the background of the Cortina d'Ampezzo start/finish line in **[48]** reveals the rugged terrain through which they have come. It seems Bertone was particularly inclined toward racing in the mountains, and his native region dictated that kind of race. This time, he was chided in *Auto Italiana* with the comment, "Less brilliant than usual was the racing of Franco Rol and Bertone."[53] He finished 5th in class and 11th overall, with a time of 04:05:26, while Giannino Marzotto took 1st place in 03:34:31. Marzotto was in a similar *barchetta* (0034M) and Bertone might have begun to realize that changing the axle ratio was a mistake.

[52] Anselmi and Boscarelli, p. 108.

[53] July 30, 1950, p. 40.

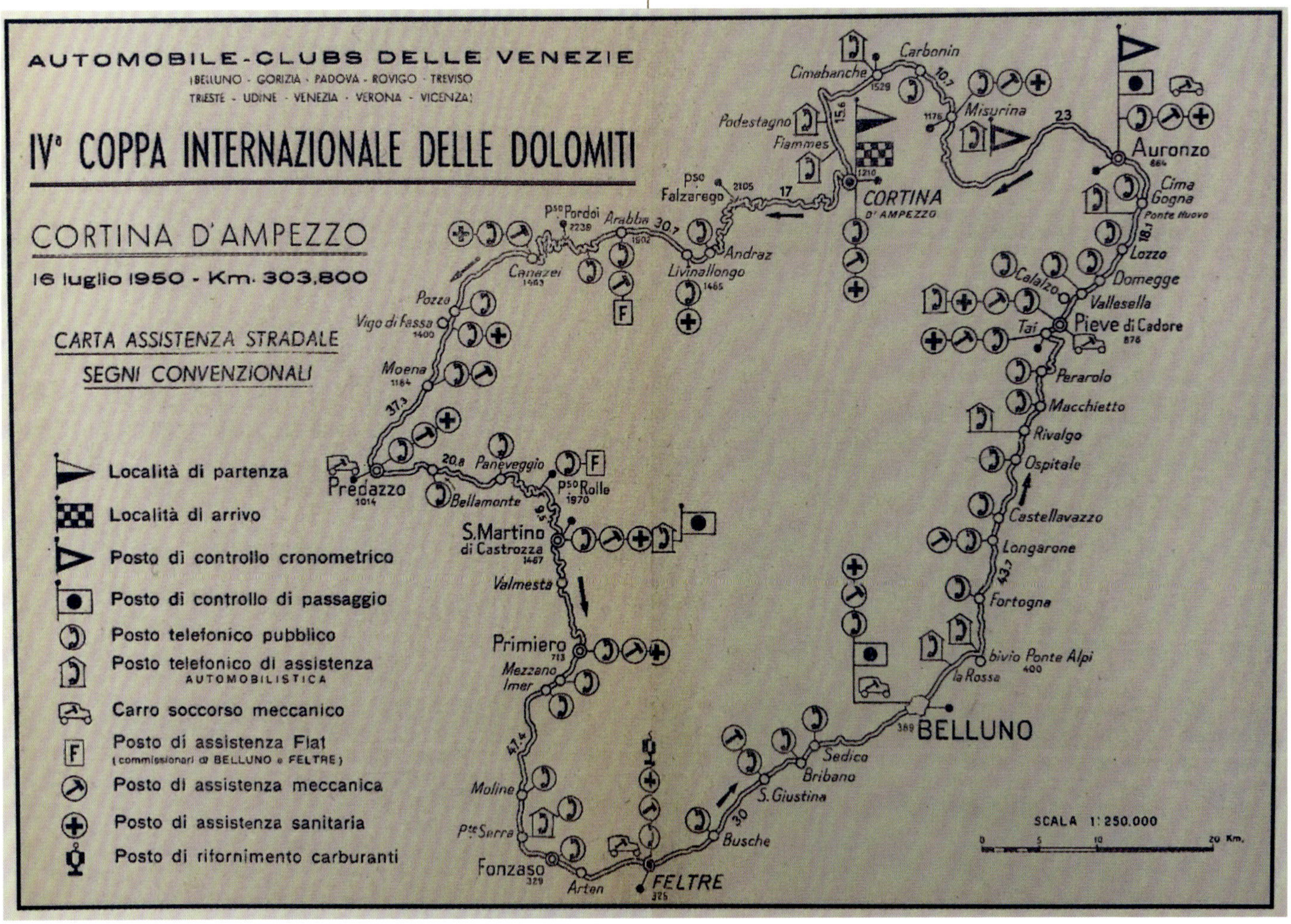

[47] IV Coppa d'Oro delle Dolomiti, 1950 – Bertone and possibly Casalegno rounding a crowded corner. *Curami Archive*

[48] IV Coppa d'Oro delle Dolomiti – crossing the finish line. *Curami Archive*

Two weeks later, on July 30, Bertone took part in the Susa-Moncenisio *corsa in salita*, or uphill race. Begun in 1902 and known as the oldest race in Italy, this was the second postwar revival. Here, drivers competed individually for the best time over 22 km from the Arch of Augustus in the town of Susa at an elevation of 660 meters up to the highest point of the Moncenisio mountain pass near the French border at 2,081 meters, a climb of 1,421 meters. In this type of hillclimb racing against the clock, the roads were often precipitous, winding and clinging to mountainsides, and their surfaces could be dangerous at speed. Courageously, drivers used a great deal of skill and nerve to balance a fast time against a fatal fall. Here Bertone placed 3rd in the over 1100 Sports class with a time of 00:16:19.1 and probably the change to a higher axle ratio had put him at a disadvantage on continuously climbing mountain roads.

A week later, on August 6, he challenged the great Aosta-Gran San Bernardo hillclimb and raced with the number 52 on the car. This event was well-described in *Auto Italiana*:

If in the case of uphill races Aosta-Gran San Bernardo is not the most beautiful race in the world, it is certainly one of the most beautiful and wonderful! The peak of G.S. Bernardo is located 33.9 km from Aosta and with a road of about 400 curves is reached from the already respectable altitude of 583 meters to that of 1,890 meters... the stretch of 10.9 km beyond S. Rhémy was in poor condition, with deep ruts, gravel, and very dusty. The belly of the engine... sometimes scraped the road, at least for certain cars, forcing them to slow down for prudence.[54]

Here, Bertone managed a more respectable 2nd in the over 1100 Sports class with the 7th best overall time of 00:27:20.4, but there were only 29 cars entered.[55] Photographs **[49]** and **[50]** show him ascending the gravel road, with the latter displaying the PROVA license plate. Interestingly, Antonio Stagnoli was 1st in class, almost one minute quicker than

[54] August 15, 1951, p. 38.
[55] www.cameva.it.

[ABOVE & OVERLEAF: 49-50] Aosta-Gran San Bernardo hillclimb, 1950 – Bertone ascending on gravel roads. *Alessandro and Massimo Acerbi, L'Aosta: Gran San Bernardo: 1920-1957*

52
PROVA
TO.99

0046M in 0018M, the first (and only other) 166 MM to have a Zagato body.

At this point, the 1950 season seems to have ended for Bertone and he might have returned 0046M to Ferrari. More and more, Carrozzeria Bertone demanded his time, and his family was becoming aware of his racing and concerned for his safety. Moreover, Bertone was not happy with his gift horse. 0046M had only a single carburetor while other Ferrari competitors' 166 MMs had been upgraded to three during the season, or even to larger engines, but then again, he was not a paying customer and one might assume that in whatever bargain he had struck with Enzo Ferrari, he was not delivering on his promise.

While Ferrari chassis 0057S was a late arrival at the Bertone works, what had already arrived was a supply of Fiat 1400 chassis, which was a popular new mass production model, and Bertone had displayed both a cabriolet and a coupe version at the Turin auto show in May. Orders for those, the Siata Amica and other volume commitments might have delayed attention to the solo chassis 0057S. As Luciano Greggio explains,

> *... the bulk of the work was concentrated between March and September, the period of the year in which both authorized dealerships and independent car showrooms were interested in purchasing and taking delivery of cars; during the other six months productivity slumped and this caused more than a few problems... not to mention the financial burden involved.*[56]

Such as Carrozzeria Bertone's finances were, they likely prioritized filling orders over the design and construction of an unsold example. Hence, the work on chassis 0057S might have been postponed until late in the year, and Enzo would not have been pleased.

While there is no proof that there was a tête-à-tête between Enzo and Bertone at this point, the Assembly Sheets for 0046M show that on March 23, 1951, the engine was converted to three Weber 36 DCF3 carburetors and the rear axle was returned to its original lower 8x40 ratio. Coincidentally or not, a month later on April 21, 1951, chassis 0057S with a Bertone body finally made its public appearance at the Turin XXXIII Salone Internazionale dell'Automobile **[51]**.

[56] Greggio, p. 138.

[51] Turin XXXIII Salone Internazionale dell'Automobile, 1951 – the Bertone Ferrari 0057S. *Revs Institute*

It seems as though Bertone was belatedly living up to an agreement with Ferrari; however, the new car looked suspiciously similar to the Fiat 1100 he had presented at the Turin show the year before, but with wire wheels. The bland, prosaic cabriolet was damned with faint praise, such as "sober," and without additional orders it slipped into obscurity.

A week after the factory completed the upgrades, Bertone took the improved 0046M and a new co-driver, Mirko Landini, to Sicily for the 1,088 km XI Giro di Sicilia race around the western half of the island on April 1, 1951, wearing number 405 **[52]**. There Bertone had finished 8th overall in 1949, driving the 1100 Stanguellini in one of the few races for which he traveled a great distance. This time he finished a dismal 7th of 8 in his class with a time of 13:31:02. He was 32nd overall, outdone by the winning Ferrari 212 Export of Vittorio Marzotto (0086E), which was followed in 2nd place by a 195 S (0038M), outclassed by another 166 MM (0034M) in 6th overall, which was followed by a 166 SC (0012M) in 9th and a 166 MM (0022M) in 10th, and

[52] XI Giro di Sicilia, 1951 – Bertone and Mirko Landini. *Boudewijn Berkhoff*

embarrassed by one of his former 1100 Stanguellinis, which finished 3rd overall.

Noticeable on the hood of 0046M in the Giro di Sicilia photograph are two small louver slits, cut to feed more air to the three new carburetors. This became a distinguishing feature of the car with no other 166 MM *barchetta* having the same, although the *berlinetta* 0060M had them as well as the 166 Inter *berlinetta* 0053S. There is also a central fog light, apparently attached through the hand crank hole in the grille.

At the end of April, Bertone and Landini entered the XVIII Mille Miglia. Figure **[53]** is another Sorlini photograph from Jess Pourret in which they are at the start line, this time pushed further back pre-dawn to 3:46 AM on the 29th, as their race number shows. Two Marchal fog lights have been added in front, replacing the central light. The young woman standing close by the left side of the car, bareheaded in the rain, could be the same one attending the Coppa della Toscana the previous year.

With 322 cars in the race, many had started ahead of them earlier in the night but they were 132nd in 02:55:23 at the end of the first stage in Ravenna (303 km), having passed many Fiats. It was in Ravenna that the photograph Bertone and Anderloni autographed for me was taken **[54]**. Soon fortune turned, however. As *Auto Italiana* reported, it was "A race initially good for Nuccio Bertone until he had the windshield broken by a stone and hammered in the face by the rain had

[53] XVIII Mille Miglia, 1951 – Bertone and Mirko Landini ready for another rainy start. *Giorgio Nada Editore Archive*

[54] XVIII Mille Miglia. Passing through Ravenna. This is the photograph autographed for me at Pebble Beach Concours, 1989, by Bertone (green ink) and Carlo Bianchi Anderloni. © *Archive Photo Locchi Florence*

to abandon at Fano, hindered because of visibility."[57] They covered about one-third of the race distance, while Paolo Marzotto in 166 Mille Miglia 0034M did very well with a 4th overall and 1st in class.

In the following month of May, a few weeks after the debut of 0057S, Enzo and Battista "Pinin" Farina met and concluded a deal between them that ended Enzo's search for "his" *carrozzeria*.[58] With that, Bertone might have given up the chase for Ferrari's business and returned the car, or perhaps more likely, Ferrari might have rescinded its use once it had been determined that his services as a coachbuilder would not be needed. Either way, Bertone raced 0046M no more after the Mille Miglia. While Enzo's reaction to the Bertone Ferrari and to Nuccio's achievements with his loaner is a matter for speculation, the fact remains that it was not until 1973, after Fiat had acquired Ferrari and controlled the production side, that Carrozzeria Bertone was given the opportunity to design the body for a series of Ferraris: the 308 GT4.[59] Enzo might have felt deceived.

In 1988, I had a telephone conversation with Luigi Chinetti, one of the closest confidants of Enzo Ferrari, and he probably summed up the situation. Talking about the scarcity of Bertone-bodied Ferraris, he said, "Very few with Bertone. We didn't like it the way he was building the body."

[57] May 15, 1951, p. 60.

[58] Luca dal Monte, *Enzo Ferrari: Power, Politics and the Making of an Automotive Empire,* p. 437. "Before the start of the 1951 F1 World Championship, Enzo Ferrari went to Tortona [to meet Pinin Farina]." The World Championship began on May 27 with the Swiss Grand Prix.

[59] Within the interval, Bertone did body two privately commissioned cars on Ferrari chassis: 250 GT short wheelbase *berlinettas* 1739GT in 1960 and 3269GT in 1962, which was Bertone's personal car.

In Bertone's defense, Luciano Greggio gives a sympathetic account of his tribulations with 0046M:

> *His most bitter disappointments were with the Ferrari. Although his '166' barchetta looked identical to the cars raced by the Marzotto brothers, it was not tuned to the same level nor did it feature the latest technical improvements as theirs did; even after he sent the car to Maranello for the necessary modifications, it was never up to the standard of the works cars, and this obliged him to fight for an honorable placing instead of outright victory.*[60]

Undeterred, Bertone continued the 1951 season racing other cars. In July, he was back in an 1100 Stanguellini, taking 3rd in class at Susa-Moncenisio, and he was at Aosta-Gran San Bernardo, finishing 3rd in class in a Maserati – on loan, of course.[61] He raced sporadically in 1952 and thereafter focused on the family business.[62]

Whatever the fate of 0046M was for the rest of 1951, it was about to receive the attention Bertone complained it had lacked.

[60] Greggio, p.125.

[61] Ibid. Loaned by Giannino Marzotto.

[62] Lorenzo Pininfarina quotes Bertone as saying in 1988 that, "I have always envied Pininfarina's relationship with Ferrari, which was consolidated over time and which has led to mutual benefits. We have established such a collaboration with Lamborghini but we have been less fortunate: that factory has always suffered a thousand hardships and so we didn't have the same constancy of relationships but short highs and lows often too long." (Associazione Italiana per la Storia dell'Automobile, *Monografia 55*) Indeed, Carrozzeria Bertone eventually closed after the death of Nuccio, while Pininfarina is still viable today.

Chapter 5

1952: GILETTI AND A CHAMPIONSHIP

0046M

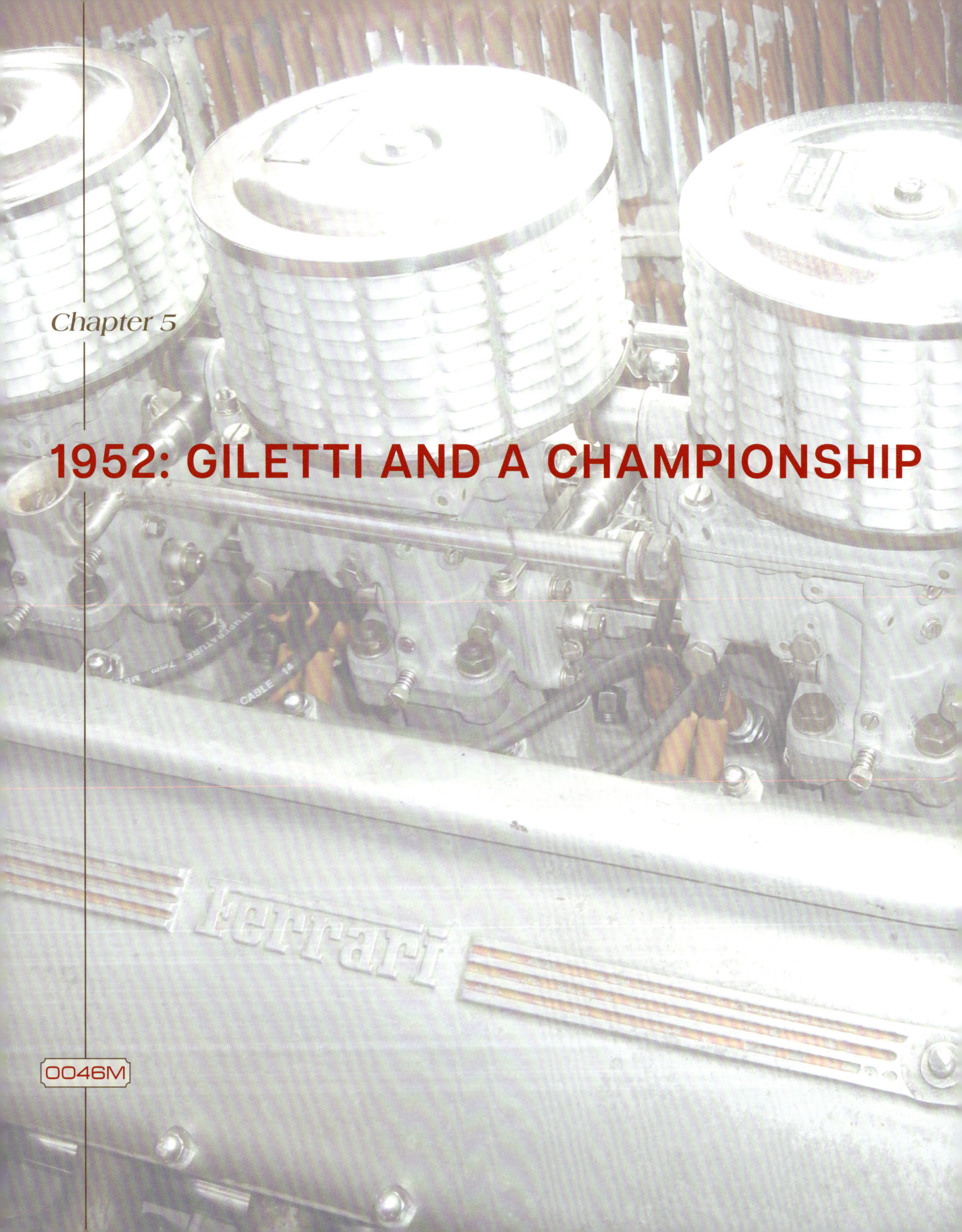

From the beginning, the 166 MM cars were modified for more performance in the quest for winning, and winning for Ferrari was not just a matter of prestige but one of survival, since winning meant sales. Early on, the 166 MM was the pinnacle of sports racing cars and ruled its 2-liter class, but competitors with larger engines soon made outright wins scarce for them. There was a cachet to class wins, but the real goal was 1st overall, besting all competition, so Ferrari was continuously building new models with bigger engines to do that.

By 1952, four years had passed since the 166 engine had been introduced and Ferrari had developed a new design by engineer Aurelio Lampredi for bigger engines and had built successful cars with it. Still, the Colombo design proved its flexibility and Ferrari had increased its power by the simple expedient of enlarging the cylinder bore to make 195 engines (2.3 liters) and 212 engines (2.5 liters).[63] Correspondingly, the same cylinder head design had been adapted for those engines by forming a shallow, concentric recess around the original 60 mm combustion chamber hemisphere of the 166 to match the larger cylinder diameters, as shown in **[55]**. Thus, an economy of scale could be accomplished by using the same stock of parts for new engines, something particularly of value since Ferrari did not establish its own foundry until 1954 and prior to that had outsourced its casting to Fonderia Calzoni in Bologna.

As the racing season dawned in 1952, Ferrari had a further development of the Colombo engine ready, the 225, which had cylinder head improvements that substantially

[63] Between 1951 and 1953, Ferrari built 21 different engines in varying use and number. Augusto Costantino, ed., *Ferrari Catalogue Raisonne 1946-1986*, p.32.

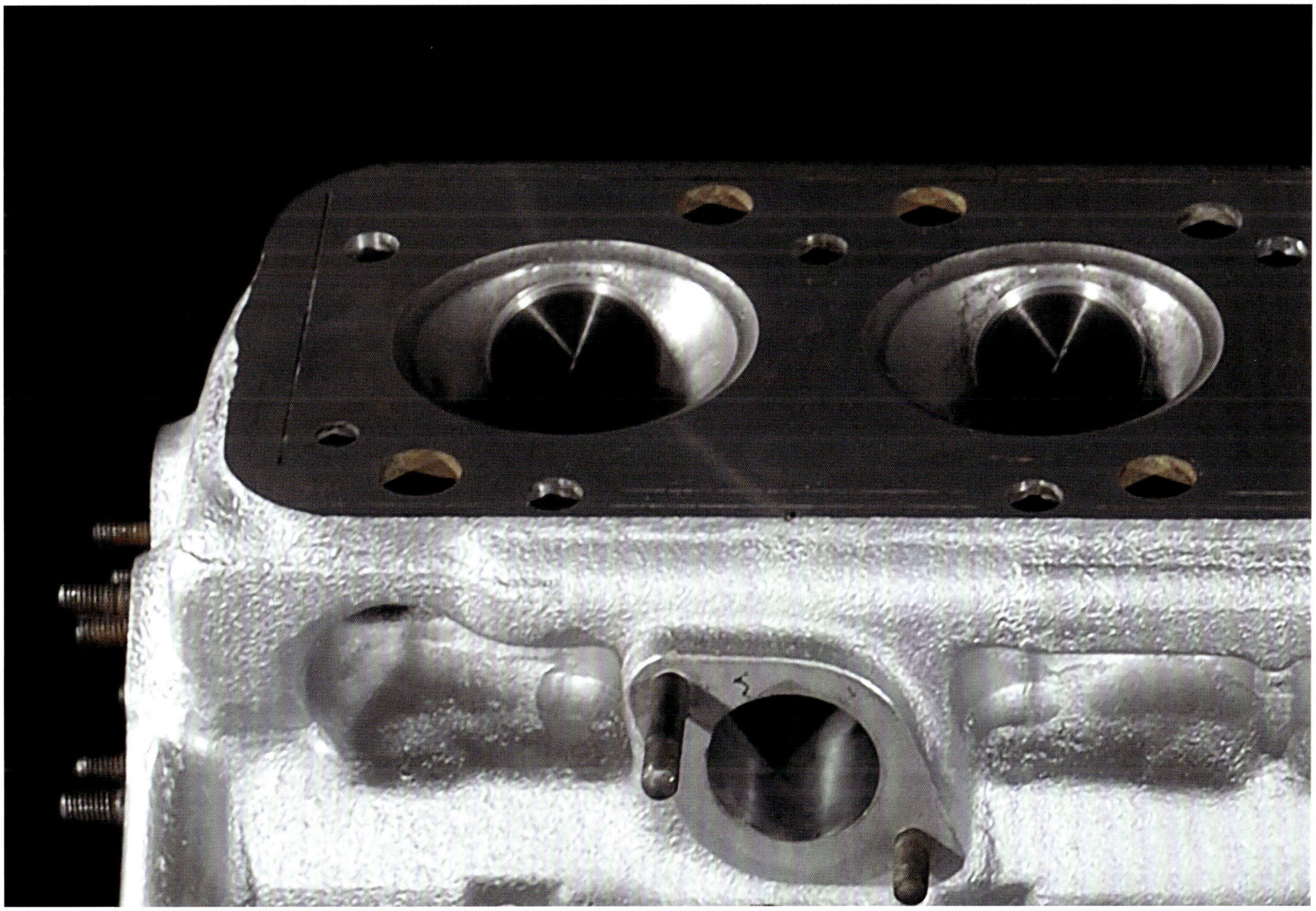

[55] 212 head with a concentric recess to match the larger bore. *Francesco Reggiani*

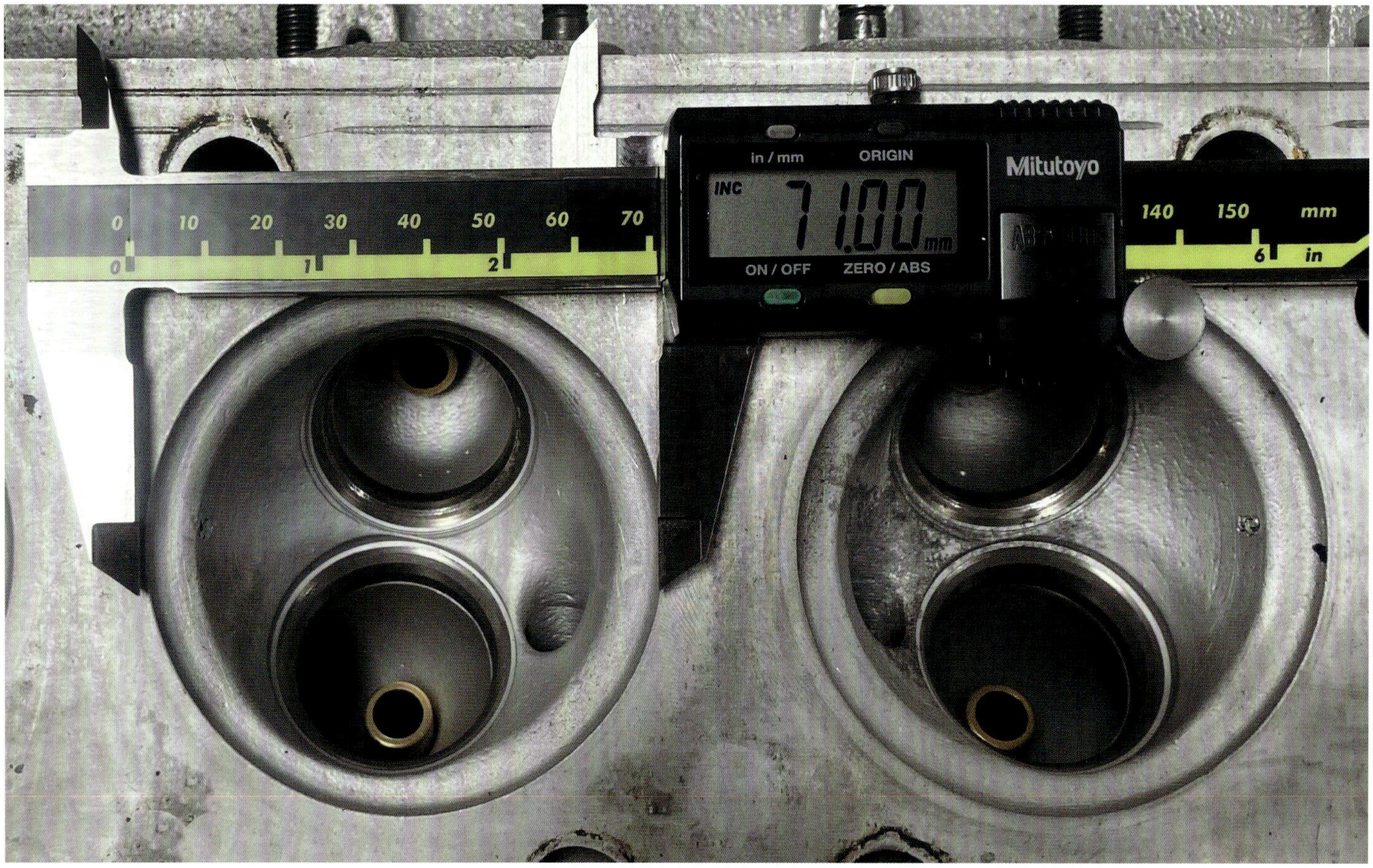

[56] 225 head with concentric recesses to match the larger bore. *Patrick and Tazio Ottis*

increased power. Since the cylinder bore was now 70 mm, the recess around the combustion chambers was widened for that and figure **[56]** illustrates how this was done on a head from a 225 engine. The 225, however, was a 2.7-liter engine and Ferrari had only the now discontinued 166 series to contest the 2-liter class in the hands of customers.

Because the 225 head followed a path of evolution from 166 heads, it was based on the 166 combustion chamber hemisphere, had the same bolt pattern and would fit perfectly on a 166 engine block if the recess around the combustion chambers was eliminated to match the 60 mm cylinders. Recognizing this, Ferrari experimented to see if such modified 225 heads could be the spark that reenergized the 166.

A few sets of the special part number 225/16658 heads were produced and two new Vignale-bodied *berlinettas* that were destined for 225 engines were commandeered for a test. 0152EL and 0164ED both had 166 engines with the new heads temporarily installed. The Assembly Sheets for 0152EL detail the progress on its intended engine, but a note says,

> *Installed engine 166 – 3 carburetors with heads and intake ducts – Experiment (with separate ducts)*

The note refers to one of the special features of the new heads – individual intake ports ("ducts") for each cylinder, giving six intake and six exhaust ports per head.

The XII Giro di Sicilia on March 9 was the first major Italian race of the 1952 season and both cars were entered as 166 MMs in the GT +1.5 class. While 0152EL did not finish,[64] 0164ED took 4th in class and 6th overall.[65] About two months later on April 4, both cars were entered in the XIX Mille Miglia,

[64] Mario Bornigia/M. Cantini.
[65] Franco Bordoni/Alberto Della Beffa.

again as 166 MMs but this time in the 2000 Sport class. Both finished in tandem with 0152EL 1st in class and 9th overall[66] and 0164ED 2nd in class and 10th overall,[67] an excellent result showing the potential of 225 heads on 166 engines for the 2-liter class.

After the two races, the test with the Vignale *berlinettas* ended and the 166 engines with their special heads were removed. Both cars were then equipped with their originally intended 225 engines and went racing with new protagonists.[68] Perhaps the new cars were considered to be better contenders for outright wins with bigger engines and as such were therefore more desirable and profitable that way.

The unlikely overall winner of the Giro di Sicilia on March 9, and winner of its class, was 166 MM number 0034M with Paolo Marzotto, the same car and driver that had finished 6th the previous year. Of the race win, Luigi Orsini notes, "As if intended to promote this category that has known better days, comes the incredible success of Paolo Marzotto... ."[69] It was the victory by five minutes of a two-year-old, 2-liter car over newer cars and bigger engines, made possible because it, too, now had the 225 heads. The Marzotto brothers had previously raced the car with great success, due to a large degree because they were favored, wealthy customers who bought Ferrari's latest cars and received the latest updates. Vittorio Marzotto had newly purchased a 225 S Vignale spyder (0154ED) to enter in the Giro di Sicilia and 0034M might have received the special heads as part of the deal.[70] Autofficina Omega S.n.c., the firm who rebuilt the engine for current owner Juan Quintano, has confirmed that it does have the 225-style heads.[71]

Following the Giro di Sicilia, the Marzottos handed 0034M to Sergio Sighinolfi with unspectacular results and then sold the car, after which it had little competitive effect. Yet, 0034M and the Vignale experiments demonstrated that the new heads could enliven the aging 166 engine.

Also racing in the Giro di Sicilia was Emilio Maria Giletti [57]. He was another young scion of an industrial family, the owners of Giletti S.p.A., a textile business founded at Ponzone in the Province of Biella in 1884 and still operating today. His ancestral estate is across the street at 39 Frazione Ponzone [58].

[57] Emilio Giletti. *Emilio Giletti*

Located in the Piedmont region of northwest Italy at

66 Antonio Brivio Sforza/Pasquale Cassini.
67 Franco Bordoni/"Geronimo."
68 0152EL had a 212 engine upgraded to 225. The Assembly Sheets note that after dynamometer tests, the 12-port heads were removed before it was installed in the car and replaced with "normal" siamesed port heads, but the instructions were to "keep the same camshafts, valve rockers, etc." Some "225" cars had the 225 valve gear but not the 12 port heads.
69 Corrado Millanta, Luigi Orsini and Franco Zagari, *Ferrari Automobili 1947-1953*, p. 253.
70 Vittorio and Giannino Marzotto entered a 225 S (0154ED) and a 340 America (0030MT) respectively, but it was only Paolo in their little 166 MM that finished, and won at that. There must have been some post-race ribbing.
71 "The engine of the Ferrari 166 MM N034 [0034M] has been modified by the manufacturer with heads having six intake ports. The valve actuation with rollers on the rocker arms is rolling eliminating the sliding action on the cams of the previous rockers. In addition, an induction system with 3 Weber 36 DCF3 double body carburetors was mounted." Private email to Laila Quintano, June 16, 2022.

[58] The Giletti estate across the street from the family textile factory. *Google*

the foothills of the Alps above Turin, Giletti, like Bertone, was based in terrain that favored hillclimb races. Again like Bertone, he hid his early forays into racing from his family. When he was old enough for a racing license at 18 in 1948, he entered the Biella-Oropa hillclimb and erased the race number on his Lancia Ardea before returning home. Early on, he raced under the pseudonym "Piliomo," a derivation from his older brother Pier Anselmo's name, and as such he entered the 1950 Mille Miglia in a Lancia Aprilia with Mirko Landini, Bertone's former co-driver, but did not finish.[72] In 1951, he showed considerable promise, beginning in a Cisitalia 202B at the Coppa Inter-Europa race where he finished 9th, then driving a Nardi for four 1sts, a 2nd and a 3rd in class in hillclimbs.

Giletti was 22 at this time in 1952 and racing was a part of his youthful exuberance. Like many young men, he sought the excitement of the big city to assert his independence and had rented an apartment in Turin at Via Pietro Micca 6. Less than a kilometer from there was a favorite haunt: Racing Club 19 at Corso Vittorio Emanuele 65.

Racing Club 19 was founded in 1948 by 19 original members, hence the name. Among them were such notables as Umberto Agnelli of Fiat, Alberto Ascari and Nuccio Bertone. This was a place where members could socialize, have a cigar, cocktail or coffee, discuss recent races and developments, or simply play cards. They were drivers, manufacturers, designers and the social elite, all bonded by the automobile culture that was thriving in Turin.[73]

Another member of Racing Club 19 was Gino Valenzano, 27, the heir to a profitable transportation business but also well-connected with the Agnellis and Gianni Lancia, which led to an invitation to race a Lancia car. For the opening race of the new season, he enlisted his club friend Emilio Giletti to co-drive with him in a 2-liter Lancia Aurelia B20 coupe at the Giro di Sicilia.

In their class, they were competing against the two 166 Vignales with 225 heads, and for an overall win they were racing against 0034M with its 225 heads. Also entered were six of the new 225 S Ferraris. Two of those were brought by Scuderia Guastalla, a premier racing team formed by Franco Cornacchia that had begun racing in January of the previous year.[74] The *scuderia*, literally a "stable", was based on Via Carlo Freguglia in Milan and was composed of both promising and proven drivers and their cars who pooled their resources and were in turn supported by an infrastructure of mechanics, transporters and logistical assistance. It was well-organized, successful and allied with Ferrari.

Franco Cornacchia's Agenzia Internazionale Commercio Auto Ricambi in Milan was the first and most important Ferrari dealership in Italy.[75] He had a close relationship with Ferrari and his Scuderia Guastalla often served to test and evaluate the latest designs, at times becoming the substitute Ferrari team. In fact, the 225 S Vignale Spyder that Cornacchia had recently bought (0162ED) was one of the first and was driven in the Giro di Sicilia by veteran Scuderia Guastalla member Giovanni Bracco but listed as a Scuderia Ferrari entry.[76]

Scuderia Guastalla's 225 S entry, the Touring-bodied car of Eugenio Castellotti and Annibale Broglia (0166ED), was the only 225 S Ferrari to finish the race, in 5th place overall. Finishing ahead of them with a smaller engine were Valenzano and Giletti in 3rd place overall and 2nd in their class. Perhaps this brought Giletti to the attention of Cornacchia, or possibly Bracco, Giletti's hometown hero, recommended him since both had won 1st in their classes at the Susa-Moncenisio hillclimb the previous year. The result was that Giletti was invited to join Scuderia Guastalla, probably as a junior member to fill in the ranks for the 2-liter class. Valenzano became a factory team member of Scuderia Lancia.

72 Luigi Perna, "La Storia Emilio Giletti," *Sportweek*, September 2019, p. 34-38.

73 *www.automobilismodepoca.it/racing-club-19-i-piloti-gentiluomini-di-torino-auto-22495*, accessed April 26, 2018, link now defunct.

74 Scuderia Guastalla was apparently named after a small town in the Emilia-Romagna region.

75 Carlo Anderloni, *Monografia 46*, Associazione Italiana per la Storia dell'Automobile.

76 *www.racingsportscars.com/results/Giro_di_Sicilia-1952-03-09-2071.html*, accessed August 6, 2021.

Later on at Racing Club 19, there must have been considerable chatter about Paolo Marzotto's unusual victory at the Giro di Sicilia with the enhanced 166 MM. Giletti, who had been chasing him during the race, surely saw the potential of the car firsthand, and since he needed a Ferrari to go racing with his new *scuderia*, would have wanted one like it. His fellow club member Nuccio Bertone, ever the salesman, might have seized on that to offer him one. Without access to Ferrari's records on the matter, I can only rely on what Giletti told me and construct a feasible scenario from that. I was able to contact him by telephone in 1990 with the help of my translator, Mary Virginia Reese, and subsequently he answered a faxed list of questions (Appendix 15). Giletti told me on the recorded telephone call that Bertone loaned him the car to try it out, that he knew Bertone from racing, and that they were good friends.

0046M was still languishing unsold with Ferrari, but the caveat to Bertone's offer of the car was that it had not been upgraded, nor had Ferrari seemed especially interested in doing so for him. However, Giletti's new team, Scuderia Guastalla, was a different matter. The situation could have been that Bertone approached Enzo with a possible sale going to his favored *scuderia*, if only the new heads were installed. To which Enzo might have countered by supplying the heads but not the labor – since he was busy preparing cars for the upcoming Mille Miglia race – and delegated the work to Scuderia Guastalla. That would have been a bonus for Enzo, since 0046M would still belong to Ferrari.

Giletti confirmed the arrangement in his faxed replies, saying,

> *The new cylinder heads were assembled for me by Ferrari.*
>
> *The installations were made by Scuderia Guastalla, behind which there was still Ferrari.*
>
> *All modifications were made by Ferrari for experience.*

The Assembly Sheets for 0046M have a handwritten entry at the bottom of the next-to-last page which is difficult to decipher, but it is a delivery note, number 474-C.[77]

[77] Identified as such by Marcel Massini, *Massini Report* for Gooding & Company, "Ferrari 166 MM Barchetta Touring 1950, Chassis #0046M."

A translation reads as follows with part numbers:

> *They have been assembled #2 heads 225/16658*
>
> *Camshafts 410/16662-63*
>
> *Roller rocker fingers 410-16641-42*
>
> *Intake valves 400-16155*
>
> *Exhaust valves 225-16676*
>
> *Valve springs 225-16670*
>
> *Rocker springs 225-16533*
>
> *Crankshaft with first undersize*
>
> [on] *main* [and] *rod* [journals]

Although it is not specified on the Assembly Sheets, this must have been the kit sent to Scuderia Guastalla for the upgrade work. As Giletti indicated, Ferrari assembled the various components, building the complex heads themselves before delivery. More than simply swapping heads, however, there was an overhaul of the engine with a reground crankshaft and new bearings, but the 225-style heads were the game changer.

As mentioned earlier, the Colombo heads were originally designed for a supercharged engine and when used on the 166 normally aspirated engines they did not "breathe" the fuel/air mixture from the carburetors well, which restricted their power. Figure **[59]** shows an early Colombo head with the intake side to the camera. While there were six cylinders for each head, there were just three "siamesed" intake ports, each feeding two cylinders and resulting in an inefficient flow. Also, the valve train can be seen, which consisted of a central camshaft (not installed in this photograph) with its lobes acting directly on rocker arms that actuated the valves, resulting in power-robbing friction and wear on the cam lobes. Aurelio Lampredi said about the problem, "The wear of the cams was an embarrassment to such an extent that we finally adopted roller rocker arms at the end of 1952."[78] Actually, the series of about 21 Type 225 cars built at the beginning of 1952 had that feature.

According to Ed Niles,

[78] Anselmi and Boscarelli, p. 36.

[59] Early Colombo 166 head showing three siamesed intake ports. *The Klemantaski Collection, Corrado Millanta*

> *The Type 225 engine was quite significant in the Ferrari family tree, for it was the first to abandon the 'shoe' type of camshaft follower in favor of the needle-bearing rollers and it was the first to use 12 individual intake ports in the heads. In fact, the engine was the direct predecessor of the 250MM and the famous 250GT with which Ferrari earned the adjective 'legendary.'*[79]

Figure **[60]** shows how the standard rocker arm cradles fit on 0046M's new heads and the new rocker arms' innovative roller bearings can be seen sitting above where the camshaft would turn in its bearing surface, while figure **[61]** provides a closer view. Each cradle carries two rocker arms actuating intake and exhaust valves on opposite sides, a robust system that reduced friction and wear. As before, there were two mousetrap-style springs beside each valve, making the assembly compact and reducing the length and therefore the weight of the valves.

[79] *Cavallino*, Vol. 1, No. 1, p. 17.

Figure **[62]** shows one of 0046M's new heads with its six intake ports, one for each cylinder and paired in flanges. On the opposite side there are six exhaust ports, unchanged from the original design. The date 14 March 1952 is stamped on both heads **[63]**.

Even though its displacement is that of a 166, the original engine block is stamped 225 under the front carburetor **[64]** while it's raised part number 166 0324 is cast in the block under the rear carburetor. Presumably, the stamping indicated modification of the block for the 225 heads by machining it for six brass inserts screwed into its top with blind threaded holes in them. Long studs that screwed into these attached the three intake manifolds, which were of a style borrowed from the engines Aurelio Lampredi had designed for Ferrari. Heretofore on Colombo engines, the intake manifolds on which the Weber carburetors sat had been bolted to the heads by studs on each side of the six siamesed intake ports, but the 225 had a solution to the

[60] Roller rocker arm cradles sitting on the 225 head of 0046M. *Author*

[61] Underside close-up of a rocker arm cradle showing details of the roller tappets and the camshaft bearing surface. *Author*

[ABOVE: 62] Left side 225 head of 0046M showing the six intake ports. The damaged chamber number ten is visible underneath. *Author*

[LEFT: 63] Date stamped on both 0046M heads, exhaust side. *Author*

[64] 225 stamped at the top of 0046M's block and the brass inserts with blind threaded holes to bolt down the new-style intake manifolds. *Author*

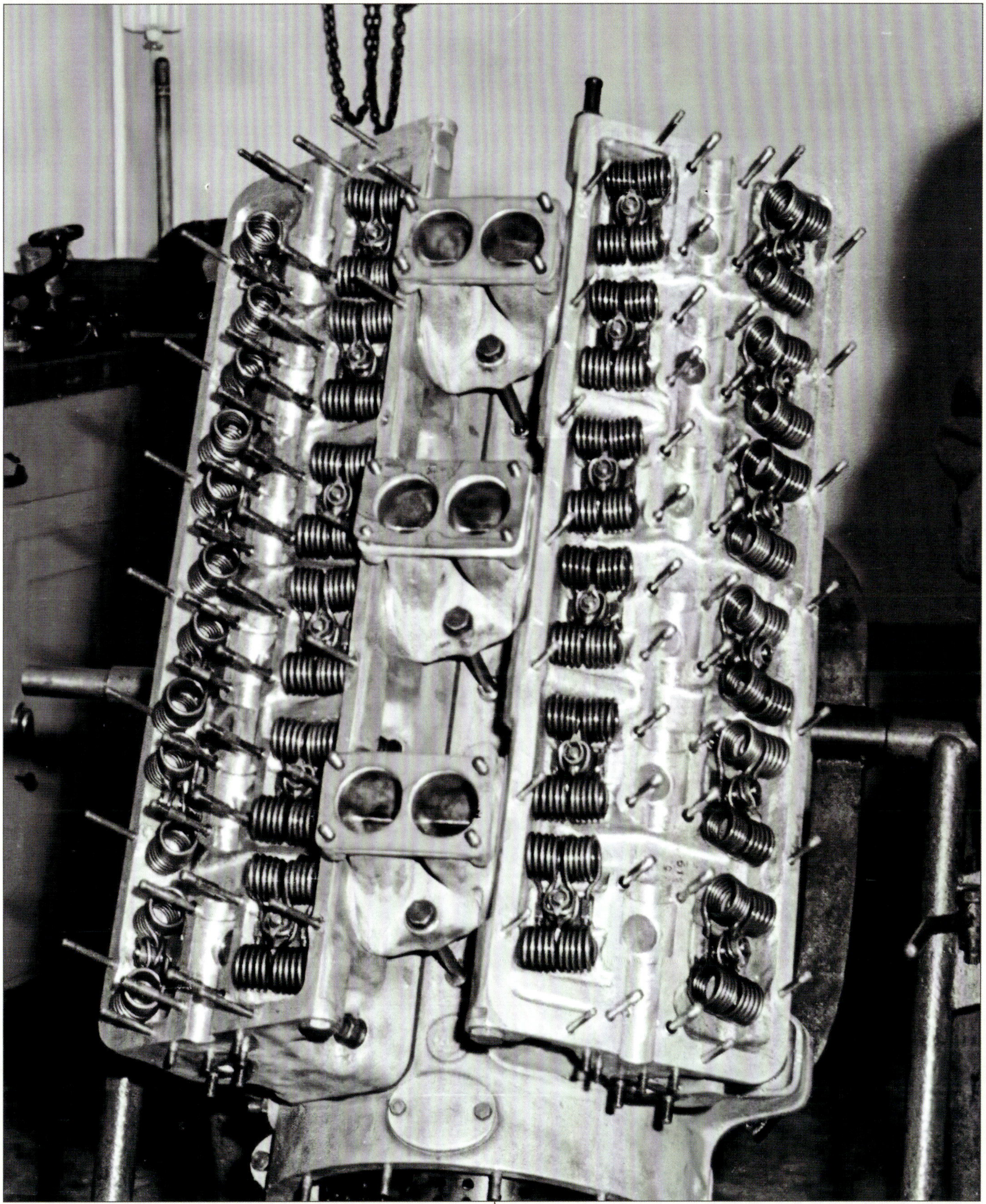

[65] A 1952 Ferrari engine photograph showing in detail how the intake manifolds are fitted. *Revs Institute*

cramped workspace in the valley between the heads where the spark plugs also were. The one-piece intake manifold for each carburetor was wedge-shaped, and when centrally bolted down between the heads on the long studs, it precisely mated against the slanted opposing intake flanges, connecting with the paired ports on each side. Inside the manifolds, the polished passages below each of the two carburetor barrels smoothly divided into two at a knife edge and those passages curved out to connect with the ports on their respective sides.[80] Figure **[65]** is a 1952 photograph of a Ferrari engine showing how the intake manifolds are similarly bolted to the block and not the heads.

Further evidence of the experimental nature of the engine are the signs of four round holes that had been cut in each of the upper corners of the camshaft covers where breathers were likely located, possibly to relieve internal pressure from the higher RPMs that the engine could now reach.[81] Assuming the covers are original to the engine, the breathers were apparently deemed unnecessary or a nuisance and were subsequently filled in with welded plugs that were polished but did not match the original metal. These can be seen later in figure **[248]** on page 230.

Judging from figures listed in subsequent Ferrari sales brochures of the period, the modifications likely resulted in an increase of 15-20 horsepower over the better 166 MM engines and extended the RPM range by 600 RPM.[82]

The delivery note for the heads on the Assembly Sheets is dated April 19, 1952, and the XIX Mille Miglia was to be run two weeks later on May 4. If Scuderia Guastalla attempted to ready 0046M for the race, there was either not enough time to complete the engine overhaul by then or problems arose, so the car was not entered.

Nevertheless, an error persisted for many years in the belief that 0046M had been raced in the 1952 Mille Miglia by Emilio Giletti. Marzotto, Prunet, Niles, and even more recent Ferrari historians attributed that to him because he did enter a 166 MM in the race and he did race 0046M that year. I did my part in promulgating the mistake by proudly repeating the current wisdom in my 1988 *Prancing Horse* article that my car had run in the 1952 Mille Miglia. When Marzotto's revised book *Red Arrows: Ferraris at the Mille Miglia* was published in 2001, 0046M was still listed as Giletti's entry, but now there was a photograph of the start clearly showing him in 0044M and not 0046M. Once that discrepancy was noticed, it cast doubt over 0046M's racing heritage and a respected website that catalogs Ferraris and their history eventually stripped all 1952 racing history from 0046M and gave it to 0044M. This remained until publicity about 0046M in 2016 revealed the error of that reaction, but the website's solution was then to credit both Ferraris with the same results! I pointed out the impossibility of this to them in 2017 and since then the duplication has been corrected in favor of 0046M.

Actually, Jess Pourret discovered back in 1990 that Giletti was in 0044M. As he searched his archives for me, he found the photograph of Giletti at the 1952 Mille Miglia start, realized the mistake, and advised me by letter how it really was (Appendix 16). I was communicating with Lynn Larson who owned 0044M at the time, and I gave him the news along with the photograph from Pourret. None of us bothered to make an announcement of the finding, so the record remained unchanged until after *Red Arrows* was published.

0044M was loaned for Giletti's use, presumably by Vincenzo Auricchio of Cremona who had purchased it the year before. According to David Burgess-Wise's research, Auricchio had previously damaged the engine and Ferrari replaced it with the engine from Formula 2 car 06C,[83] so it would have been a worthy substitute. Photographs show it carried the Scuderia Guastalla emblem on the fender tops, indicating Giletti was a team member at that time and Cornacchia's influence probably played into the loan. Giletti had just

[80] Jumping ahead, the 166 MM/53 series that followed 0046M had four-barrel Weber carburetors and a further improved intake manifold with an independent passageway for each intake port, one from each carburetor barrel.

[81] The Assembly Sheets for 0152EL note that on the first running-in test of its early 225 engine there was a "Strong oil leak from the flywheel side, from the generator drive and from the tachometer drive."

[82] Anselmi and Boscarelli, p. 104.

[83] Private history for Sir Michael Kadoorie, current owner of 0044M.

turned 23 and was joined by Walter Loro-Piana whom Giletti described in his scrapbook as his "good friend and mechanic" and who might have been from the fellow textile business family of that name.[84] Unfortunately, they did not finish the race, going off the road only 40 km from the start.

[84] The Loro-Piana family is famous for its brand of cashmere wool.

Giletti's next race was the IV Coppa della Toscana on June 1, and this time he really was in 0046M, along with Loro-Piana again. He certainly knew what the significance of 0046M was. As he told me,

> *In addition to being a racer, I had a great passion for mechanics and also served as a tester, collaborating on the testing of changes or new applications which, if the results were positive, would then be used for production. In 1952 we tested a different cylinder head with success, which beginning in 1953 was also used for other engines.*

Photographs **[66]** and **[67]** show Giletti in Livorno standing beside 0046M on the right before the Coppa della Toscana and wearing his usual colorful socks. Loro-Piana is on the left in his familiar leather jacket in spite of the summer heat. The photographs before the start of the race also show a man talking first with Loro-Piana and then with Giletti. He resembles Bruno Moroni, a Scuderia Guastalla member not entered in the race but who knew the course from the 1950 event and who was familiar with the 166 MM from his own 0048M *berlinetta* version, all of which would make his advice valuable. Important to note is the license plate on the Ferrari, M 23742: it is a cardboard temporary license issued for a motor vehicle to operate. These were similar to the "dealer" tags seen today and the "M" suggests Cornacchia's Milan dealership. No records were kept of those, but it is a further indication that 0046M had been a car of unregistered ownership to this time, operating previously under Bertone's PROVA TO 99 license as a manufacturer's test car.

An official wields his stamp on the way to the start line in **[68]** and the race number, 1240, again indicates the start time. At the starting line **[69** and **70]**, the aforementioned man can be seen from two angles leaning over behind Giletti and looking at the instruments with concern after the recent rebuild of the engine. When the car is flagged off **[71** and **72]**, he has moved only a short distance away on the right and still appears to be watching the instruments like a hawk. Moroni, or whoever he was, was very much involved. Later,

[66] IV Coppa della Toscana, 1952 – Giletti stands to the right side of 0046M while Walter Loro-Piana (leather jacket) is being coached on the left side. © *Archive Photo Locchi Florence*

[67] IV Coppa della Toscana – Giletti receives his last-minute instructions. © *Archive Photo Locchi Florence*

[68] IV Coppa della Toscana – An official stamps their time card on the way to the start. © *Archive Photo Locchi Florence*

[69] IV Coppa della Toscana – Consultation at the starting line. © *Archive Photo Locchi Florence*

[70] IV Coppa della Toscana – Waiting for 12:40. © *Archive Photo Locchi Florence*

[THIS PAGE: 71 & 72] IV Coppa della Toscana – Flagged away. © *Archive Photo Locchi Florence*

[73] IV Coppa della Toscana – Charging on after the Arezzo control point. © *Archive Photo Locchi Florence*

they are passing through a control point at Arezzo **[73]**.

Ultimately, it was a splendidly successful debut for the new heads: Giletti won 1st in class and placed 11th overall out of 213 cars starting the 736 km race. His time of 06:28:15 was 10 minutes ahead of the 2nd place in class Maserati and *Auto Italiana* called it "a quiet victory" for Giletti with nothing special to report in the 2000 Sport race.[85] Since only 121 cars finished, it must have been more than uneventful for almost half of the entrants.

Undoubtedly, Giletti was impressed with the power of the Ferrari's radically improved engine compared to that of 0044M he had raced several weeks before and, seeing its potential for the rest of the season, must have wanted the car. 0046M, though, had always been in the inventory of Ferrari, which could have affected its sale. By 1952, Ferrari had begun to establish a dealer network to which, historian Luigi Orsini explained, "the 'normal' customer was invited to turn. But especially in the early years, for the 'special' and/or wealthy customer... the doors of Maranello were open."[86] Even so, as Luigi Chinetti told me,

> *A few times, the factory did deal direct. One case was the King of Belgium – but I was getting the commission too – and the Morocco king. But Ferrari protected always the dealer, but never made the deal. The customer had no* [way] *to do the deal direct with the factory but always through somebody.*[87]

While such dealer protection may have held true for touring

[85] June 15, 1952, p. 47.

[86] Private email, July 25, 2020.

[87] Chinetti related a colorful story about Enzo to me. He was explaining how Ferrari sold cars through his dealers except in special cases and said, "Ferrari he asked me long ago when he met Sophia Loren, he say, 'Luigi, I like to sell a car to Sophia Loren.' A beautiful woman he like to deal direct to him himself even if the customer come from us. About woman he has very good taste, Ferrari. That's why he make a beautiful car too."

cars and the "normal" customer, the sale of competition cars was likely more circumspect and overseen by Ferrari to ensure success. Bertone, though, was equally qualified to both buy and retail 0046M, first having been entrusted to race the car and then having become a de facto Ferrari dealer with 0057S. And so, the *Torino Ufficio del Pubblico Registro Automobilistico* (Appendix 3) records that he finally bought the car from Ferrari on June 9, 1952, but he did not register it for a *licenza circolazione*, or license plate to allow its use on the road, because he was buying it for resale, and he may have sold it to Giletti on the same day.

A few days later on June 12, Giletti raced the car again in the IV Giro dell'Umbria (Coppa della Perugina), a revived series that had last been run in 1927. A photographic aberration makes 0046M appear a light color in **[74]** where Giletti and Loro-Piana are seen settling into car 304 prior to the race. In the next photograph **[75]**, the Scuderia Guastalla decal **[76]** is clearly seen on top of the fender.

[74] IV Giro dell'Umbria, 1952 – Giletti speaks to someone as Loro-Piana climbs in. *Anonymous*

[75] IV Giro dell'Umbria – Giletti looks confident as they wait to start. *Anonymous*

[76] The Scuderia Guastalla emblem on the fender of 0046M. *Scuderia Guastalla Le Nostre Vittore 1951, 1952*

[77] IV Giro dell'Umbria – Giletti slides the car around a corner on gravel while Loro-Piana looks back to see who follows. *Barchetta.cc*

There is a movie of this race on *YouTube* and at 00:03:31 there is a momentary view of 0046M's nose as it sits before the race, revealing the distinctive two hood slits and the capped hand crank hole at the bottom of the grille, but it is incorrectly identified as Carlo Gazzabini's number 308 (0074E ?). Then at 00:04:42 there is a brief look at the car during the race and shows the white paper license still at the rear.[88] In **[77]**, Giletti and Loro-Piana are rounding a corner at speed and here the darker shade is realistic for the car's color. The race distance was 383 km and Giletti finished with a time of 03:16:25, which was six minutes behind the winner. Once more 0046M proved itself in his hands, finishing 1st in class again and 3rd overall, ahead of two larger 2.5-liter 212 Ferraris.[89]

It was not until June 27 the Turin Public Register recorded that the license plate TO 134705 had been assigned to 0046M on June 26 and showed the sale from Bertone to Giletti, giving his apartment address in Turin. In a 2019 interview for a magazine article, the 90-year-old Giletti recalled that he signed promissory notes for two years.[90] The price was 3,600,000 lire, or about $5,760 in 1952 dollars.[91] Converting that to 2024 dollars, the price was about $68,600, which would be a bargain for a two-year-old Ferrari today. I remember seeing an article long ago in one of my brother's old 1950s hot rod magazines with a picture of a new Ferrari that listed the astounding price of $12,000.

Of course, Franco Cornacchia with whom Giletti was allied in Scuderia Guastalla was a major Ferrari dealer, so the question arises, why did he not buy the car through Cornacchia? For one thing, as Giletti said, he and Bertone were friends, and both were members of Racing Club 19. Simply put, Giletti might have favored his "fraternity brother" who offered the car first. Another factor to consider is Ferrari's sense of loyalty to his dealers, as Chinetti described above. Bertone had a long-standing association with 0046M as a "demonstrator" car, so perhaps he had the first option for its sale, or argued that he did.

Bertone raced sporadically in the 1952 season and then, bowing to the wishes of his family, he retired to lead the family business to fame as an innovative designer and to fortune as a manufacturer.

Under the auspices of Scuderia Guastalla, Giletti was able to expand his racing activity all over Italy, whereas being a privateer the previous year he had mostly confined himself to the Northern provinces closer to home. On June 29 he entered the XXXVI Targa Florio in Sicily, for that year a race consisting of 8 laps around a mountainous portion of the island, each one 72 km with 898 turns. In **[78]**, Giletti and number 70 enter the complex of shaded grandstands at

[88] *www.youtube.com/watch?v=gVxmGXUpYU0*, accessed April 29, 2018.

[89] Prisco Giuseppe, *Coppa della Perugina: giro automobilistico dell' Umbria*, pp. 184-185.

[90] Perna, p. 38.

[91] *www.wikipedia.org/wiki/Italian_lira*, accessed January 17, 2018. Subsequent lira to dollar conversions are based on the 1949-1970 rate set at $1=625 lire.

the start/finish line in Cerda dubbed "Floriopolis." Festively decorated with garlands and bunting, the stands are packed with spectators watching for a brief glimpse of the cars as they make their infrequent passage. Having rounded the turn after the stands, Giletti passes the raised pit area on his left and is on his way back to the countryside **[79]**.

There is a movie of that race which has been posted on *YouTube* and at 00:01:02, car number 70 can be seen coming in for a pit stop and then leaving.[92] It was here, though, that Giletti met his first disappointment with 0046M. He was running 5th overall on the 2nd lap but began losing ground, about a minute each lap, indicating that the car was ailing. Indeed, the race movie shows that the hood was

[92] *www.youtube.com/watch?v=CnlSdrTcnVM*, accessed January 20, 2018.

[79] XXXVI Targa Florio – Giletti passes the pit area on his left. *Targapedia.com, Archivio Storico Peppino Cappellani*

[78] XXXVI Targa Florio, 1952 – Giletti enters the Floriopolis complex. *Fotografia Randazzo*

up during the pit stop. He fell to 6th on the 3rd lap, 7th on the 4th lap, 9th by the 5th lap, then moved up to 8th on the 6th lap because of a retirement, and on the penultimate 7th lap he himself retired.[93] But the Targa Florio was a notoriously hard race and only 15 cars finished out of 47 at the start.[94]

Skipping the month of July, the next race Giletti entered was the IV Giro della Calabria on August 3, and he was given the race number 702, but either did not arrive or did not run.[95] The layover of one month might suggest that repairs had been necessary after the Targa Florio.

By the next weekend, August 10, Giletti and 0046M were ready to resume the fight at the VIII Circuito Automobilistico di Senigallia race. This was the culmination of the III Settimana Motoristica delle Marche (3rd Motoring Week of the Marches, the central Adriatic coast of Italy) and followed motorcycle races the previous weekend. The Senigallia course was 9.3 km and regarded as the second fastest in Italy after Monza, having a 4 km straight, wide curves and sound pavement. It was a true test of the power, braking and handling of a car and the skill of its driver. For the race, classes of 750 cc, 1100 cc and over 1100 cc had been set, but there were so many entries for the over 1100 cc class that the *Commissione Sportiva* divided it into 1100-2000 cc and over 2000 cc, which benefitted Giletti by giving him manageable competition.

The 2-liter class was contested by 10 cars in a 15 lap race. At the start, Inico Bernabei's Fiat 8V took the lead, followed by Enzo Pinzero's Ferrari 166 (0022M ?), Diego Capelli's Fiat 8V, Eugenio Castellotti's 166 MM (0058M), and Giletti (number 15). On the second lap, Pinzero set the lap record for the class and took over 1st when Bernabei dropped out with fuel pump failure. By the 5th lap, Castellotti had moved to the front, followed by Pinzero, Capelli and Giletti in 4th. On the 14th lap, Giletti outpaced Capelli, equaling the lap record at 00:03:53.4 and taking over 3rd place to the finish.[96] Thirty-eight years later, he was still proud of that record and pointed it out to me. His total time was 01:00:10.3 for the race.

With the withdrawal of Alfa Romeo for 1952, the Formula 1 World Championship had been downgraded to Formula 2 cars to fill the grids, and consequently several courses changed their scheduled Formula 1 event to a sports car race. At Pescara, a six-hour race had been held with good success before the Coppa Acerbo Formula 1 race the previous year. Dropping Formula 1 and mounting a more ambitious event this time, the XXI Circuito di Pescara race became the 12 Ore di Pescara, a 12-hour endurance race for sports cars on August 15, a grueling test for 0046M.

For a race of that duration, a relief driver was needed, so Walter Loro-Piana, being a mechanic as Giletti described him, was left on the sideline and Giletti partnered with Sergio Sighinolfi, a promising 26-year-old who was a veteran

[93] *Auto Italiana*, July 15, 1952, p. 32.
[94] *Giornale di Sicilia*, June 30, 1952.
[95] *Interauto*, September 1952.
[96] Benito Quadraroli, *Circuito di Senigallia*, p. 189.

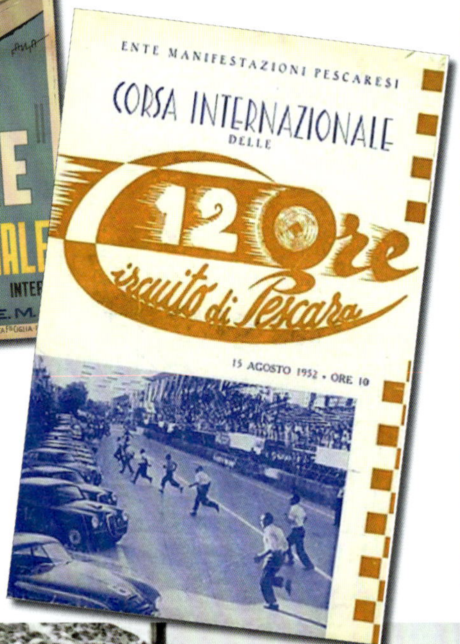

of four years racing and had finished his final race for the Marzottos in 0034M with its similar engine two weeks before. Each lap was 25 km through and around the city of Pescara on sometimes winding, sometimes blindingly fast, streets and rural roads that were often lined with crowds, which made it a difficult and dangerous race. Adding to the element of danger was the Le Mans start in the north central part of the city, where the drivers were lined up along the old Via Nazionale Adriatica on the opposite side from their 42 diagonally parked cars. At the 10 AM start signal, they raced across the road, jumped into their cars, started their engines, and charged out into the mêlée, hopefully avoiding a catastrophe.

In the first photograph, **[80]**, the drivers are sprinting for their cars then all are seen getting aboard **[81].** 0046M is well down the line, the first car on the far side of the checkered finish line and notable by the single driving light below the grille. The cars are roaring off in **[82]** where 0046M can be

[80] 12 Ore di Pescara, 1952 – Le Mans start. 0046M is just behind the finish line with a driving light mounted at the bottom of the grille.
Monica Valeriani

[81] 12 Ore di Pescara – Drivers reach their cars and climb in. Giletti is opening his door. *Monica Valeriani*

[82] 12 Ore di Pescara – Forty-two cars charge out into the street. Giletti is above car 7. *Monica Valeriani*

[83] 12 Ore di Pescara – Giletti (18) looks for an opening as the Uovo Ferrari (6) comes out. *Luigi Orsini*

[84] 12 Ore di Pescara – Screen capture from a movie showing a bird's-eye view of [83] with Giletti indicated by the arrow. *Archivio Luce*

seen through the space behind front row car number 7. In [83] the famous "Uovo," or egg, 166/212 MM 024MB (6) of Guido Mancini/Fabrizio Serena is pulling out as Giletti (18) follows Luigi Piotti/Vittorugo Mallucci in 212/225 Ferrari 0104E (9).

Archivio Storico Instituto Luce has a film of the start and the chaos is amazing to watch from above.[97] A screenshot at 00:13:59:16 [84] is the same moment the previous photograph was taken, and puts in perspective where the cars are, indicating Giletti with an arrow. At 00:14:53:15 in the video, the expanse of the racecourse is plain as Giletti races down a long country road while in [85] there is urban racing at the "S" bend in Cappelle sul Tavo with Giletti being led by Diego Capelli in a Fiat 8V Zagato. We can see the single central driving light that was mounted low in the center of the grille for when night came. This was probably attached with a temporary clamp, either through the slats of the grille or through the hand crank access hole.

Because he was grouped with the over 2-liter cars, Giletti placed 4th in that class behind three larger Ferraris and was 6th overall, covering a distance of 1,405.269 km. 1100 cc OSCA MT4s were 4th and 5th. The winner traveled a distance of 1,539.833 km.

---

97 *www.patrimonio.archivioluce.com/luce-web/detail/IL5000042112/2/gara-automobilistica-durata-pescara.*

[85] 12 Ore di Pescara – Diego Capelli leads Giletti at Cappelle sul Tavo. *Monica Valeriani*

In his faxed reply to me, Giletti listed racing at Enna on August 31, saying he finished in 3rd place and added Giro di Sicilia in parenthesis. Enna was the site of the Grand Prix of Pergusa, Circuito della Cravatta, in Sicily on that date and period reports do not mention him there. It is possible that he was thinking of his 3rd place finish in the Giro di Sicilia on March 9. However, one deceased historian noted in his records that Giletti did not finish the Enna race and that his car number was 17.[98] Antoine Prunet, also now deceased, listed Giletti at Enna in the records he gave me. Perhaps he was at least present.

Some sources list Giletti at a Trofeo Supercortemaggiore race in Merano, Italy, on September 13, where he supposedly was 1st in class. However, it seems this is another case of repeated misinformation. Merano is in the extreme north of Italy, near the Austrian border, and it is unlikely that the Ferrari could have been transported by the next morning from there to where Giletti and 0046M definitely were.

Early on September 14, Giletti was 357 km off the west coast of Italy on the island of Sardinia, and with him again was Walter Loro-Piana. It was the second running of the Trofeo Sardo, a race up and down the island, beginning at Cagliari at the south end, up to Sassari in the north, then after a rest stop, back to Cagliari again, a total distance of 420 km. There were 60 cars entered and 8 in the 1100-2000 cc class with Giletti. To his advantage, the organizers excluded cars above 2000 cc. After his 9:04 AM staggered start, but with the car simply numbered 10, Giletti led at Sassari where the cars were lined up in order while the drivers took a break before

[98] Private email from the person in possession of said archive.

[87] Il Trofeo Sardo – Carabinieri and even a priest seem amused as Giletti and Loro-Piana are perhaps trying to read a map. *Salvatore Marras, Archivio Privato Mario Pintus, Sassari*

[86] Il Trofeo Sardo, 1952 – Giletti and Loro-Piana arrive in first place on Viale Acquedotto in Sassari. *Salvatore Marras, Archivio Privato Mario Pintus, Sassari*

[88] Il Trofeo Sardo – 0046M positioned for the restart on Via Serra Secca as Giletti has a conversation and Loro-Piana relaxes. *Salvatore Marras, Archivio Privato Mario Pintus, Sassari*

[89] Il Trofeo Sardo – Loro-Piana and Giletti face the camera with the cars lined up on Via Serra Secca. *Salvatore Marras, Archivio Privato Mario Pintus, Sassari*

restarting for the return run. Photographs [86-89] were taken in Sassari showing Giletti arriving on Viale Acquedotto and then parked ahead of Diego Capelli's Fiat 8V Zagato on Via Serra Secca, awaiting the restart. Loro-Piana leans against a rear fender while Giletti first leans against a tree in conversation, probably with Capelli, then stands to the other side of the car facing the camera. At the finish in Cagliari, he took 1st overall with a time of 03:03:16.9, about two minutes ahead of Capelli. Winning the race earned Giletti the prize of 450,000 lire – 150,000 for his class win and 300,000 for the overall win[99] – or about $720 in 1952 dollars, which was about $8,600 in 2024 dollars. Worth more, perhaps, was the recognition an outright win brought.

His next race was on September 28 at the VI Gran Premio di Bari on the Adriatic coast of Italy. Like Pescara, the race was to have been for Formula 1 but the Automobile Club of Bari likewise decided to make it a sports car event. Ferrari, of course, held the edge there as well, and of the 28 cars entered, 19 were Ferraris.[100] Prior to the race, the cars were assembled in a piazza for numbering with 0046M the third car on the right with 73 [90].

There is a movie of this race, too, that is posted on *YouTube*[101] and the field can be seen roaring away from a standing start with Giletti appearing on the right moving away from the camera at 00:00:21. At 00:00:29 he is seen closer, moving toward the camera on the left with the hood slits plainly visible while the number 73 is there but difficult to discern. Giletti's characteristic white driving suit, his black helmet, and even his mustache can be distinguished if viewed frame by frame.

[99] Nello Alfano, *La Cagliari-Sassari-Cagliari*, pp. 43-72.
[100] Luigi Amati, *50 anni di automobilismo sportivo In Puglia*, p. 80.
[101] *www.youtube.com/watch?v=RKJ-kO80eHw*, accessed January 18, 2018.

[90] VI Gran Premio di Bari, 1952 – Cars gather in a piazza for numbering. 0046M is 73, the third on the right. *The Klemantaski Collection, Corrado Millanta*

From the starting line on the old Lungomare della Fiera road beside the Porto Nuovo Harbor, the course traveled counterclockwise, curving around the edge of the peninsula along the Adriatic Sea where Giletti leads Alberico Cacciari's 166 Export 0072E rebodied by Carrozzeria Campana in **[91]**. After two turns and a straight heading inland, the cars turned left onto what is now Via Napoli, a long, straight road with the highest speeds of the course. At about the middle, Giletti has passed under the pedestrian bridge at the intersection with Via Ammiraglio Caracciolo **[92]**. From Via Napoli, cars braked hard to make a sharp left turn onto Via Brigata Regina, and Giletti leads Sergio Mantovani in 166 MM 0058M at that corner in **[93]**. A rough repair to the bodywork of 0046M is visible on the front right fender

[91] VI Gran Premio di Bari – Giletti leads a rebodied 166 along the Adriatic Sea. *Luigi Orsini*

[92] VI Gran Premio di Bari – Speeding along the Via Napoli straight, having passed under the pedestrian bridge. *The Klemantaski Collection, Corrado Millanta*

[93] VI Gran Premio di Bari – Giletti leads Sergio Mantovani at the end of Via Napoli. *The Klemantaski Collection, Corrado Millanta*

behind the headlight, indicating there must have been a collision previous to this race. In **[94]**, Giletti has opened up his lead over Mantovani and there is a perspective of the corner and the length of Via Napoli, while the next photograph **[95]** shows him exiting that corner onto Via Brigata Regina. That short street connected by a curve with Lungomare della Fiera road at the harbor to complete the 6 km lap. Photographs **[96, 97 and 98]** were taken during and after the curve. As these images show, the numerous rail lines serving the port[102] were an obstacle Giletti had to

[102] Today, the rail lines have been removed and roadways changed since the Port of Bari converted to trucking.

[94] VI Gran Premio di Bari – Another lap on Via Napoli showing the length of the straight. *The Klemantaski Collection, Corrado Millanta*

[95] VI Gran Premio di Bari – Exiting Via Napoli onto Via Brigata Regina. *The Klemantaski Collection, Corrado Millanta*

73
73
TO·13
4705

73
134703-TO

[OPPOSITE & ABOVE: 96-98] VI Gran Premio di Bari – Sequence showing the curve from Via Brigata Regina onto Lungomare della Fiera. *The Klemantaski Collection, Corrado Millanta*

cross repeatedly and might have been the reason he had a tire puncture,[103] costing time and resulting in an 11th overall finish, which was only achieved because 12 cars dropped out. Even so, he was only 1½ minutes behind the winner, finishing in 01:08:24.3, a gallant effort which earned a 5th in class as well. It was a good showing for the Ferraris with 225 engines since they filled the first five places.

The upcoming photograph of 0046M in the *1952 Ferrari Yearbook* **[102]**, (seen on page 112) shows Giletti (320) at the III Bologna-Raticosa hillclimb two weeks later on October 12. On the front right fender can be seen the repairs noticeable at Bari. Running from Bologna to the Raticosa Pass, halfway to Florence, this was the second longest hillclimb in Europe at 43 km and climbed 348 meters to an altitude of 520 meters. Here, Giletti was 3rd in class and 5th overall with a time of 00:28:25.4, which was 45.8 seconds slower than the overall winner.[104] The photograph at the finish **[99]** shows him met by Loro-Piana (at left, with his hand on the car behind Giletti) and who appears to be Franco Cornacchia (at right, the large man with his left hand on his hip). He may be explaining to Cornacchia the new damage along the right side and dent in the driver's door, evidence of contact with something solid that impeded his progress.

For the final race of the season, the leaders of Racing Club 19 proposed to Scuderia Sant Ambroeus of Milan that its Turin team be allowed to join the Milan team's concluding race as a contest between cities, a thoroughly Italian tradition. Scuderia Sant Ambroeus was a racing team similar to Scuderia Guastalla, but without the blessing of such a

[103] *Ferrari Automobili 1947-1953*, p. 301. In **[96]**, the left side latch for the trunk is open, perhaps left that way in haste after the tire change.

[104] *Auto Italiana*, October 15-30, 1952, pp. 59-60.

[99] Bologna-Raticosa hillclimb, 1952 – There is a dent in the driver's door after the run. *Boudewijn Berkhoff*

patron as Ferrari, and which had been organized by Elio Zagato, son of Ugo Zagato, founder of Carrozzeria Zagato, and his friends. At the end of 1951, they had celebrated the conclusion of their first year with the Coppa Sant Ambroeus race at Monza, which was run under a handicap formula that allowed Elio Zagato to win 1st overall with a 750 cc car. For the 1952 race, however, the formula was changed to accommodate Racing Club 19 and reward the potential of more potent cars.[105]

With that, Giletti and other Racing Club 19 members ended the 1952 season on October 26 at the II Coppa Sant Ambroeus. When the Monza track was constructed in 1922, it was the world's third permanent race course. Originally, it was a high speed oval with two high banked curves

[100] II Coppa Sant Ambroeus, 1952 – Rounding the Curva del Porfido at Monza. *Emilio Giletti*

[105] Ibid., pp. 63-65.

connected by straights, but it evolved in stages to eliminate the high banking and incorporate more curves and straights. In 1952, the track length was 6.3 km. In **[100],** 0046M as number 200 is rounding what appears to be one of the consecutive south curves, known as Curva del Porfido, on the old surface of square porphyry cobblestones, which in a later revision became the Curva Parabolica. Giletti recalled the race:

> *I participated in a race in Monza in 1952. I was forced out of the race due to the engine on the 12th lap* [of 15], *after I set the best lap time, however, and while I was leading the race.*

Giletti's recollection coincides with the II Coppa Sant Ambroeus report in *Auto Italiana.*[106] Regardless of that shortcoming, his accomplishments with 0046M throughout the season earned him enough points to win the Championship of Italy for the 2000 Sport class. As reported in *Auto Italiana*, the *Commissione Sportiva Automobilistica Italiana* ranked Giletti 1st in his class with 20 points. Furthermore, he was ranked 3rd in *Assoluto Sport* (all Sport classes) with 12 points in that category.[107]

Giletti was honored by Racing Club 19 at their annual banquet and presented a trophy by the club's president, Giorgio Giusti, seen in a worn photograph from Giletti's personal scrapbook **[101]**.

Another image of the moment was published in the magazine *Motor* with the caption:

[106] Ibid.

[107] *Auto Italiana*, November 30, 1952.

[101] Racing Club 19 annual banquet. Giletti receives the Fiat Cup as Champion of Italy, 2000 Sport class. *Emilio Giletti*

EMILIO GILETTI Il giovane industriale di Ponzone Biellese, pur essendo entrato soltanto quest'anno nel campo delle vetture di media cilindrata, ha meritatamente conquistato il titolo di campione italiano della Categoria Sport, classe 2000, con vettura Ferrari 166.

Oltre al Trofeo Sardo, che ha vittoriosamente coronato la sua brillante stagione, Giletti ha fatto spicco nei Giri della Toscana e dell'Umbria, sui circuiti di Senigallia e di Pescara, nella Bologna - Raticosa. È probabile che il forte pilota piemontese passi alle maggiori cilindrate nel corso della stagione che sta per cominciare.

> *Dr. Giorgio Giusti proceeded to the end of the year awards of the Turin sportsmen, during a convivial gathering attended by more than one hundred persons. There were many telegrams of support, including that of our newspaper and our director Favia. In the photograph, Giorgio Giusti delivers the Fiat Cup to the Italian champion Emilio Giletti. At the symposium there were also elegant ladies and a wide representation of the Italian industry.*[108]

In addition, Giletti was lauded in a Racing Club 19 magazine article, which was captioned "Our Champion" and led by the photograph in **[100]** (shown on page 110). Noting his successes and pointing out that he had only just begun, it commented:

> *It is too early now to say that Giletti is an ace in the broadest meaning of the term. But it cannot be denied that the ascending parabola of our friend can lead him to the highest goals.*[109]

In further recognition of Giletti's success, Ferrari devoted a full page to him in their *1952 Ferrari Yearbook* **[102]**. The text reads as follows:

> *The young industrialist of Ponzone Biellese, despite having entered only this year in the medium displacement class, has deservedly won the title of Italian champion of the Category Sport, Class 2000, with the Ferrari 166.*
>
> *In addition to the Sardinian trophy, which has victoriously crowned his brilliant season, Giletti has been prominent in the tours of Tuscany and Umbria, on the circuits of Senigallia and of Pescara, and in the Bologna-Raticosa. It is probable that the fast Piedmontese driver will move to the major displacement class during the next season.*

[108] *Motor*, Giletti clipping, no date.
[109] Giletti clipping, no date.

[OPPOSITE: 102] The 1952 Ferrari Yearbook recognized Giletti's success and included a photograph from the Bologna-Raticosa hillclimb. *Ferrari S.p.A.*

Despite the praise of the *Yearbook*, Giletti once said that he went to Modena to visit Enzo Ferrari and asked him if he was glad that he had won the championship. Instead of praise, his reply was: "But look, that is not your credit. You won because of my car." Giletti went on to say, "He did that with all the drivers. I confirmed that with Alberto Ascari."[110]

To a degree, though, what Enzo Ferrari told Giletti was true. The year 1952 had been one of tremendous success for Ferrari and Enzo had been named a Knight of Industry in Italy, but overall, 166 MMs had not fared well. As Anselmi and Boscarelli put it,

> *The 166s were on many occasions decisively beaten by the Aurelia, the Fiat 8V, the Gordini, and also by the nimble OSCA Tipo MT4... The competitive qualities of the no-longer young 166 were now confirmed as inadequate; this, combined with the great popularity of the 2000 cc sports car class, persuaded Ferrari to introduce for the 1953 season an updated version of the car, the 166 MM/53.*[111]

However, 0046M with its special heads was a different 166 MM in 1952 – the shining exception – and Ferrari, having taken note, used the same heads to power the 1953 series of 166 MM cars. I believe it can be said that in Giletti's hands, 0046M played an important role in advancing the evolution of Ferrari and that it has been overlooked for having done so.

[110] Perna, p. 39.
[111] Anselmi and Boscarelli, p. 140.

Chapter 6

1953: BOSISIO AND ZAGATO

0046M

Giletti was moving on, having seen the writing on the wall from the new 2-liter Maseratis, and so he was willing to listen when 28-year-old Luigi Attilio Bosisio of Milan wanted to buy the Ferrari. I asked Giletti why he sold it, and his reply was simply that Bosisio made him a good offer. At the Milan *Ufficio del Pubblico Registro Automobilistico* (Appendix 4), the sale price was recorded as only 100,000 lire, equal to $160 then and about $1,900 in 2024, so either there was a digit missing or there must have been some private consideration as well. The "good offer" Bosisio made could have been to pay token equity and then assume the balance of the two-year note Giletti signed. Tired from a year of hard racing, 0046M may have succumbed to wear and tear from the last race at the Coppa Sant Ambroeus, and whatever the price, it may have reflected that work needed to be done. Nevertheless, the championship winner still had much life in her yet.

Giletti's success brought him a spot on the Maserati factory team for 1953, and regardless of the *Ferrari Yearbook*'s prediction of his moving to bigger engines, it was the 2-liter A6GCS sports car he raced with best results, such as 1st in class and 6th overall in the Mille Miglia, 2nd overall at the Targa Florio, and a transatlantic 6th at the 1,000 Kilometers of Buenos Aires. He became good friends with famed teammate Juan Manuel Fangio, entertaining him at his family estate in Ponzone and besting him in the 1953 Targa Florio. In 1955, at age 26, he retired from racing at the insistence of his father to run the family business, as Bertone had done.

Bosisio had been employed since 1942 at age 18 by Pirola Editore, a Milan publishing company dating back to 1781. Undoubtedly, he was a high achiever since he went on to become the owner of the company, but his hobbies were far from bookish. He flew airplanes and by 1953 he was moving in the moneyed circles of amateur racers.[112] His 1954 racing license is reproduced above **[103]**, but he began racing in 1952 in a loaned Lancia Aurelia and soon joined Elio Zagato in the Scuderia Sant Ambroeus, which was based at 21 Viale

Fédération Internationale de l'Automobile
COMMISSIONE SPORTIVA AUTOMOBILISTICA ITALIANA
ROMA Via Po, 14 MILANO Corso Venezia, 43
N. 0268
LICENZA DI CONDUTTORE
VALEVOLE FINO AL 31 DICEMBRE 1954
Rilasciata al Sig. BOSISIO LUIGI
pseudonimo autorizzato
nato il
domiciliato a Milano, Conservatorio 28/15
li 11.3.1954
IL TITOLARE
Il Presidente della Commissione Sportiva Automobilistica Italiana
Automobile Club d'Italia

[103] Luigi Bosisio's 1954 racing license. *Luigi Bosisio*

Luigi Majno, in Milan. For his new team, he needed a new car and what better than a championship winner? Bosisio already knew the car and knew Giletti. As described earlier, I was able to reach Bosisio by telephone in 1989 and interview him through my interpreter; he told me,

> *When he* [Giletti] *raced with the Ferrari barchetta I used to race with the Venti's Lancia. So we raced together and we knew each other really well.*

The Public Register lists the sale date as January 3, 1953, with the address listed for Bosisio as Via Conservatorio 28/13 in Milan, an apartment building. That was recorded on January 10 in Milan, but it seems Bosisio took possession of 0046M soon after the II Coppa Sant Ambroeus and already had the intention of converting the car to a *berlinetta*. He said,

> *When I bought the car it was still in the Touring style, but I changed it right away. Because I bought it around October or November and I brought it to the body shop right away and we changed everything... we changed everything that was done by Touring. We changed the front of the car, we put the roof on that didn't exist before. We changed the doors and there wasn't much left of what Touring did.*

Bosisio explained his reasoning for converting 0046M to a *berlinetta*, saying,

[112] *Corriere della Sera*, "A questi quattro illustri milanese l'insegna di cavaliere del lavoro," June 3, 1985, and "Pirola vuole creare un nuovo quotidiano economic," June 5, 1985.

> *It excelled in the long races; I never raced short races; I have always run races that were hundreds of kilometers long. It was a closed car and that was a big advantage because it is less tiring. It does not get wet inside, even when it is raining. You get less tired and you can last longer.*

The body shop was Carrozzeria Zagato at Via Giorgini 18 in Milan and the photograph below **[104]** shows the facility in 1947 with Elio Zagato on the left and his father Ugo in the center. According to Michele Marchianò in his book *Ferrari by Zagato*, work began in December after consultation with friend and teammate, Elio.[113]

Ugo Zagato had started his own *carrozzeria* in 1919 after spending years of apprenticeship in automobile construction and then gaining experience building airplane components during World War I. He applied techniques from the latter to automobiles, pioneering the use of strong but lightweight materials along with streamlining in an era of heavy, boxy vehicles. The advantage in performance as well as appearance was noteworthy, attracting orders through the 1930s on such chassis as Alfa Romeo, Lancia and Fiat. World War II interrupted progress, but when business could be resumed, Zagato had a new development: Engineer Fabio Luigi Rapi collaborated with Zagato to patent a unique style known as "panoramica." Inspired by the vista from airplane canopies to make use of the light, pliable new material Plexiglass, Zagato produced several early examples in 1947 giving the cabins a "greenhouse" effect with windows along the sides that curved into the roof at the top and also curved toward the rear of the car, creating an aerodynamic, teardrop shape. In some cases, a shaped panel of Plexiglass above the windshield expanded the forward view.

Influences of earlier Zagato designs can be seen in 0046M: the 1949 Maserati A6 1500 Panoramica for example **[105-106]**, and especially the Ferrari *berlinetta* built carte

[113] p. 56.

[104] Carrozzeria Zagato facility, 1947. Elio Zagato left, his father Ugo Zagato center. *Archivio Zagato ©*

[105-106] 1949 Maserati AG 1500 Panoramica with curving Plexiglass windows and windshield extension. *105: Tim Scott, 106: Archivio Zagato ©*

blanche in 1949 on chassis 0018M for Antonio Stagnoli, the Piaggio Vespa scooter dealer in Milan **[107-108]**. The Ferrari exemplified Zagato style with its curving side windows and uninterrupted contour of the body. In the only angular feature on the car, the windshield was a split one of two flat glass panes, while the top tapered down gracefully at the back and blended into the roundness carried back from the rear fenders, similar to 0046M. A few photographs are all that remain of the panoramica body on the 0018M chassis since Stagnoli had Zagato change the body to a cycle-fendered spyder after a year, and although the car has survived, it is said to be in very bad condition.[114] It is safe to say, then, that 0046M has Zagato's earliest bodywork remaining intact on a Ferrari and that it is certainly the earliest Ferrari remaining with panoramica influences.

By the early '50s, a muted panoramica style had evolved as the trademark of Zagato, now less radical in the upper

[114] In 2007, Zagato built a replica of the 0018M *berlinetta* for a client on the chassis of a 166 Vignale Ferrari.

[107] Antonio Stagnoli with his 166 MM, 0018M, the first Zagato Ferrari. *Archivio Zagato ©*

[108] 0018M displays Panoramica hallmarks. *Archivio Zagato ©*

curvature of the side windows and the notion of windshield extensions discarded. Still, the dictum was that wind and weight were the enemy of performance, so the typical Zagato design was rounded, tapered, smooth and as light as possible.

Marchianò suggests 0046M has a stylistic inheritance from Zagato's first custom body on a Fiat 8V chassis built for Ovidio Capelli, the Milanese Fiat dealer, in 1952, citing its rear window shape.[115] Frankly, the rear windows are different, but that 8VZ does have a center-rear fuel filler as 0046M has, although it is below the rear window, and a second car, a roadster, built for Capelli by Zagato the next year had the same feature. When Zagato began a series production of 8VZ cars in 1954, the fuel filler location was frequently centered under the rear window beneath a small door.

Another influence on 0046M could have been the body Zagato built for Luigi Piotti on OSCA[116] chassis 4502 as modified by Gian Paolo Volpini from a 1951 Tipo G Grand Prix car with one of three 4.5-liter V12 Grand Prix engines **[109-110]**. Although that body no longer exists on the car and a few photographs are the only evidence of it, there was a family resemblance with 0046M between the greenhouses, both having reduced side window curvature and split windshields, but there was more. In its single-seat Grand Prix configuration, the OSCA had the fuel filler on the center line of the car. On the new Zagato body, the filler tube extended through the center of the rear window **[110]**, which is a signature feature of 0046M. Retaining the vertical filler tube of the Grand Prix car's fuel tank, the OSCA had its filler cap level to the ground and at a tangent to the window, while the tube on 0046M is angled and the cap is parallel to the window, perhaps showing an evolution. The date Zagato began work on the OSCA is vague, but some sources indicate it was in 1952 and thus preceded 0046M.[117]

[115] Ovidio Capelli was the uncle of Diego Capelli.
[116] *Officine Specializzate Costruzione Automobili* – the second company founded by the Maserati brothers.
[117] Luigi Orsini and Franco Zagari, *OSCA: la rivincita dei Maserati*, p. 53, and carrozzieri-italiani.com/listing/osca-v12, accessed March 5, 2025.

[109] Zagato OSCA with two windshield wipers ahead of the driver. *Archivio Zagato ©*

[110] Zagato OSCA chassis 4502 with the fuel filler barely visible through the rear window below center. *carrozzieri-italiani.com*

Since Zagato was modifying an existing body with 0046M and not starting with a bare chassis, some thought and planning was required. As Bosisio explained,

> *With Zagato we redid all the body of the car. That year I was helped by Professor* [Mario] *Speluzzi.*[118] *He is dead now. He was a famous professor from the Polytechnic*

[118] Mario Speluzzi was also in charge of pre-race scrutineering of cars for the *Commissione Sportiva Automobilistica Italiana* and as such Elio Zagato often ran afoul of him. (Elio Zagato, *Elio Zagato, storie di corse e non solo*, p. 84.) Speluzzi was also the author of several books on the technical aspects of automobiles which might have made for a mutual attraction with Bosisio in the publishing business.

> *University of Milan. He suggested the aerodynamic lines. He collaborated with the company Zagato because they wanted to create a style that was very new. For that time, it was a very modern style.*

Before work began, the engine was removed, evidently needing repair after the II Coppa Sant Ambroeus race. Bosisio said,

> *It was disassembled and fixed.* [Carlo] *Canavesi took care of the engine. He oversaw the mechanical part. Canavesi was a chief test driver of Alfa Romeo.*

When work got underway, Zagato eliminated the distinctive "omega" shape of the Touring *barchetta* nose, which had a crease over the grille extending halfway down each side and then out under the headlights and was emphasized with a scalloped area between the headlights and grille. All of that was smoothed out, presenting an uncluttered face to the wind, and whereas the nose originally projected outward at the bottom, like a plow, it was pulled back to make the grille stand upright, creating a more aggressive look. Changing the angle of the grille also meant that the hand crank access hole in the grille then pointed downward, no longer aligned with the hole through the radiator, which probably made it difficult to turn the engine manually for timing the distributors **[111]**. Also, the bottoms of the fenders below the headlights were raised slightly in order to increase air flow to the finned drum brakes, which needed all the cooling they could get. Turn signal lights were added on each side

[111] Zagato pulled the bottom of the nose back, which angled the handcrank hole downward. *Author*

as well. By the Trofeo Sardo, Giletti had removed the fairing beneath the nose that covered the bottom of the radiator, possibly for more cooling, but Zagato replaced it with a much sturdier and more protective steel cage with holes for air passage. These changes and more can be seen in my following photographs as well as others in this chapter from Bosisio's personal album and those that he provided to Marchianò for *Ferrari by Zagato*.

Along the sides of the body, below the waistline and between the wheel wells, the only change was the addition of a forward-facing flip-out vent door on each side to bring air into the foot wells, something sorely needed in a closed car.

Above the waistline running along the sides was where the major work took place, while nothing was changed on the front fenders except for the area around the headlights, which were "Frenched" into the fenders without the chrome trim rings. On the hood, the only change was the removal on each side of the small Touring badge and the *Superleggera* script beneath it. The two slits at the front of the hood remained, which Marchianò incorrectly stated were a Zagato addition after the removal of a hood scoop that was never there, and the hood was not lowered as he said either, its height being dictated by the engine, radiator and frame structure underneath.[119]

The cowl behind the hood was adapted at its back edge

[119] *Ferrari by Zagato*, p. 56.

[112] Interior view of the framework added for the windshield. *Author*

to meet the two flat pieces of safety glass that formed the windshield, the bottom edges of which were cut to match the curvature of the cowl, and on each side A-pillars were put on in addition to the center post between the panes **[112]**. While the angularity of the flat, split windshield does look incongruous against the smooth curves of the rest of the body, it was nevertheless a practical concession.

The size and shape of the windshield panes evidently caused some consternation with the placement of wipers, since **[113]** shows a third wiper arm (without a blade) on the driver's side, mounted closer to the center post. A single wiper in that position would leave too much area unswept, so Zagato probably tried to find a way to have two wipers working together. I assume they were trying to find the optimum way for wipers to sweep the windshield but ultimately decided there was none and the inner wiper was abandoned and left inert. I can attest that the driver's vision in the rain is minimal. At the back of the car in **[113]**, Bosisio is wearing a light color sweater as the race number is being painted on for the V Coppa Toscana.

Zagato also tried the use of two wipers on the aforementioned OSCA, which had two wiper posts close together in front of the driver, and in **[109** and **110]** (seen on page 119) a smaller auxiliary wiper on the right of a main wiper can barely be seen. Evidently, that experiment continued on 0046M. Later on, Zagato built some 8VZ cars with two wipers on the driver's side that were more successful on a one-piece curved windshield.

In conjunction with the wipers, windshield washer nozzles were installed on the cowl on either side of the center post and were operated by a manual push-pull pump which drew fluid from a reservoir under the hood, the bracket for which is visible later in **[168]** (seen on page 170). The pump was under the dash to the right of the passenger's horizontal grab bar visible in **[17]** (seen on page 28) and the wiper two-speed pull switch was on the central panel below the dash. No doubt the thinking was that the co-driver would operate the washers and wipers while the driver concentrated on racing.

[113] A third wiper arm was tried in front of the driver then abandoned. *Luigi Bosisio*

The tops of the doors above the waistline were changed considerably from the low-cut curve of the *barchetta* to a higher, horizontal line. This ran across from the bottoms of the windshield A-pillars to meet a line on the reshaped rear fenders that angled slightly downward toward the rear. The line across the doors and rear fenders formed the base for the distinctive Zagato panoramica side windows.

While the frames of the doors were being modified, mounting plates were attached beneath the aluminum skin for exterior door handles needed on the closed car. Slightly above the waistline, the retractable door handles helped keep the shape aerodynamic. These were a standard Zagato component made of aluminum for lightness and consisted of a lever that remained in a flush recess until the button-shape on the short side of the lever's pivot was pressed. That would cause the long, handle-shaped side of the lever to pop out where it could be grasped and pulled, allowing the other end to depress and operate the lock mechanism inside the door. Cleverly, Zagato positioned the devices so that they engaged the existing levers that operated the original latches **[114** and **115]**.

Each door had a fixed window and an adjustable vent window. The inside tops of the doors below the windows were dressed with a polished aluminum plate, visible later in **[141]** (see page 141). A practical feature was a concave depression in the plates below the thumbscrews that locked the vent windows in position. This gave clearance for fingers to turn the screw.

Frames to mate with the tops of the new curved door windows and rear quarter windows were welded over each side with B-pillars that curved at the top to match them; on those, the rear quarter windows were attached with hinges **[116]**.

More framework was added to support the top. A sturdy, semi-oval hoop was welded on the existing framework behind the rear windows and another curved span connected the B-pillars. Both preserved the torsional rigidity of the chassis and formed the basis of the roof structure **[116** and **117]**. Fitting tightly over that form, the aluminum top gently rolled down at the sides to meet the curvature of the windows and had a slight crease over the windshield center post. Since the tops of the doors were inboard of the bottoms, a nice touch was the addition of small rain gutters on each side for protection against drips in inclement weather.

At the rear of the car, the trunk lid was discarded and the rear fenders were joined across the back in a smooth, rounded surface that curved both in toward the center and up from under the car to merge with the steeply down-sloping top. For rear vision, a convexly curved Plexiglass back window

[114] Exterior retracting door handle. *Author*

[115] Interior door latch mechanism. *Author*

[116] Framework added for the top, interior side view. *Author*

[117] Framework added for the top, interior rear view. *Author*

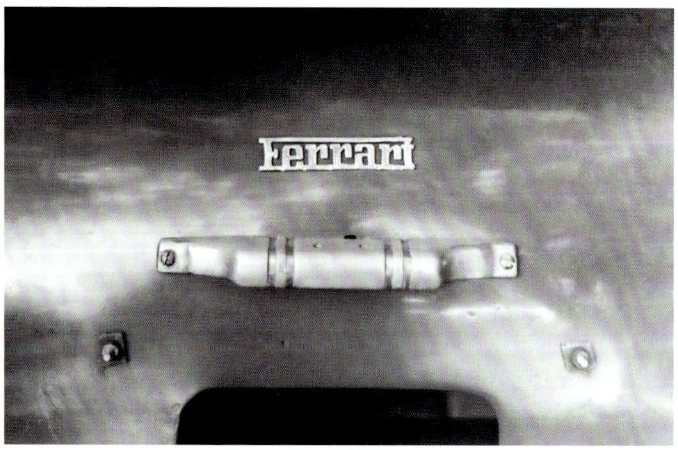

[118-119] The access hole for the enclosed rear, light fixture for the license plate and Ferrari logo. *Author*

was set low into the top, its braced frame further contributing to the strength of the structure. Through the bottom of the rear window a wide filler tube passed, going down to the fuel tank below, and was capped on the outside by a polished aluminum quick-release cap. The cap was likely the same one that had been on the car, and it is marked VETT-125 on the inside, probably the same part that had been in use since the 125 series. The space between the tube and the window was sealed with a thick ring of what appears to have been a putty-like material, which can be seen later in **[141]** (seen on page 141) along with the following details.

Beneath the filler cap, on the body, Zagato affixed its name in small chrome plated letters. Further down below the rear window, centered between the two taillights, was a rectangular access hole through which could be reached the fastening and wiring of the taillights, since the rear of the car was now completely enclosed. Covering the access hole was the license plate in a frame; over that was a polished aluminum light fixture; above that and taking the place of the original Touring badge was the Ferrari name in its iconic logo **[118 and 119]**.

Within the enclosed rear was the fuel tank. As a later generation 166 MM, 0046M was constructed having a rectangular tank inside the trunk with soldered seams, whereas early cars had tall, narrow tanks with riveted seams, rounded tops and the filler cap at the top center, a holdover from the first Ferraris. In the early configuration, the spare tire was located on a slant behind the fuel tank and facing the trunk opening. In the later configuration, the spare tire was carried on a platform above the fuel tank with a short filler tube and cap off to the right side, still accessible only by opening the trunk lid. The latter arrangement can be seen on 0052M **[120]**, the second *barchetta* following 0046M and whose Touring body number is 3449.[120]

The Assembly Sheets for 0046M specify that it was built with a 120-liter (31.7-gallon) fuel tank. The rear of the tank ended above the major cross member at the end of the

[120] Touring body number 3448 was 0050M. 0048M was a *berlinetta* and out of production sequence.

[120] Inside the trunk of 0052M. The spare tire is above the fuel tank with the filler cap off to the right. The battery has been relocated. *Rainer Wolf*

[121] Inside the enclosed rear showing the access hole, framework and one of the tubes added to mount the new fuel tank. *Author*

chassis with its back "feet" bolted to brackets on that, while three small tubes of body support framework extended rearward from the cross member. This can be seen in the trunk of 0052M. A panel across the back of the aluminum housing in which the tank rested and another over its top enveloped the tank.

At some point, the original tank was replaced with a longer one of larger capacity to extend the Ferrari's range for endurance races. The new tank had its filler tube centered at the rear. For the change, the rear and top panels of the tank box were removed, then two thick tubes were welded in from the chassis rear cross member to the aft framework on each side to support the new tank; interior photograph **[120]** shows one of them. The position of the new rear mounts on those tubes was moved about half the distance of the available free space, which would allow the new tank to carry perhaps 150 liters, an extra 30 liters or eight gallons of fuel. With so much extra weight in fuel at the back of the car, inverted U-shaped brackets made of thick sheet steel bent at right angles were bolted to the frame on each side over the axle tubes near the springs to limit upward travel of the rear axle.

Since the tank change is not noted on the Assembly Sheets, it must have occurred after 0046M left Ferrari's hands.[121] The movie of Giletti's pit stop at the Targa Florio on June 29, 1952, shows the car being refueled and the filler is definitely on the right side then, deep in the trunk, as original. Neither Giletti nor Bosisio mentioned changing the tank, but then they were not asked, which leaves its origin open to speculation.

The simplest explanation is that Zagato procured and installed a larger capacity tank during the renovation when the rear of the car was opened up for new bodywork. They had previously built the Ferrari 0018M, and that car had the early-style tank with a centered fuel filler, but it was deep in the car and awkwardly accessed by opening the rear window hatch. No doubt there were fumes in the car after fueling, hence the ventilation louvers in the rear window, and perhaps that led Stagnoli to have Zagato rebody the car as an open-air spyder after a year.[122] Both 0018M and 0046M had the same chassis, and the bodies would be similar, so a tank with its filler opening located at center rear might have been intentionally sought to avoid the ventilation problem

[121] Had the tank been installed in the *barchetta* body, its filler tube and cap would have needed to be positioned on the right side due to space limitations.

[122] Contemporary photos of the spyder version of 0018M show the fuel filler cap located at center, close behind the cockpit, confirming that the chassis retained its early-style fuel tank.

by copying the solution used on the Zagato OSCA: put the filler cap outside through the rear window. The external center position had been used by Zagato before on several cars and on designs of the time by Vignale and others, thus it followed a practical trend allowing equidistant fueling from either side during pit stops and it avoided the ungainly external side filler cap Touring used on some *berlinettas* in favor of a neat, symmetrical design.

The new tank, however, was evidently not a new one at all. Like the original tank, the current one has a deep u-shaped indentation running from top to bottom at the forward end for clearance from the differential, and the housing in which the tank sits has a corresponding indentation that slots into the tank. However, the replacement tank has damage on each side of the indentation that appears to have been caused by being struck at mid-height, as though a differential was driven back into it, but there is no corresponding damage to the tank housing or any sign of repair, showing that it did not occur in 0046M **[122-124]**. Did the tank come from a wrecked car? Whatever its source, the tank had its filler opening located at the center rear and that integrated with the design of the new body.

In another interior modification, Zagato cut out 0046M's original rear bulkhead of tubular framing behind the seats, down to the enclosure of the rear axle, leaving what had been the trunk open to the interior of the car. The existing angle iron frame on each side above the fuel tank then supported a 13 mm-thick, 88 cm-wide and 80.5 cm-deep plywood platform having a 13 cm-square notch at the back edge for the filler tube. With the increased length of the new

[122] The fuel tank of 0046M after I removed it with difficulty from the car. *Author*

[123] Each side of the tank's recess for differential clearance has damage. *Author*

[124] Tank housing showing no sign of damage and repair. *Author*

tank and the space opened behind the seats, the spare tire could now fit centered on the platform ahead of the filler tube and leather straps were used to hold it in place, visible later in **[187]** (see page 187). There was then a means of accessing the spare tire with the rear enclosed because it could be reached from inside the car, over the seats, as **[125]** shows.

Further forward in the interior, the original dashboard had been recessed underneath a frame tube at the back of the cowl that curved in three dimensions: downward from the highest point on the cowl and back at the sides to meet the curve on the tops of the doors. This defined the cockpit, seen in **[67]** (page 89) and **[75]** (page 94). Eliminating the open cockpit, Zagato replaced the curving tube with one straight across at the back of the cowl that simply arched between the bases of the A-pillars. A new dashboard was fashioned to wrap over it and match the angled base of the windshield.

The new dashboard was a rather complex piece of hand workmanship, and its width of 1,200 mm indicates how small the interior of the car is. And there is a mystery about it, too: Underneath the black leather facing that was on the dash when I bought 0046M, and clearly apparent from the back, I found holes filled in with welds showing a different and rather unusual arrangement of the instruments from the original Touring *barchetta* pattern evident in **[67]** (see page 89). The tachometer had been placed to the right of the steering column and the large round cluster of oil pressure,

[125] Spare tire access became over the seats. *Author*

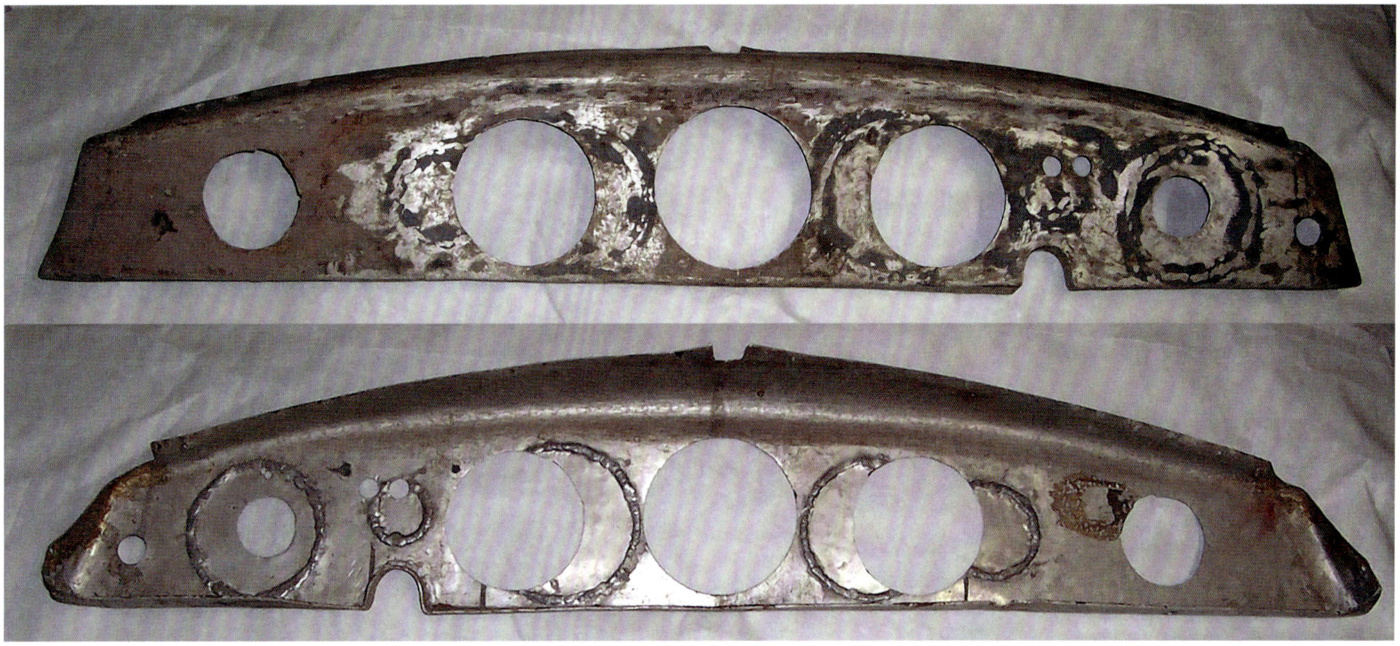

[126] Front and back of the new dashboard with welded plugs from an earlier instrument arrangement. *Author*

water temperature and ammeter instruments had been on the left of the column. A second oil pressure gauge had been added above the steering column, while the speedometer and a new instrument had been situated close together in front of the passenger, as **[126]** (bottom) reveals.

A second oil pressure gauge is not present with Giletti in **[67]** (page 89) at the IV Coppa della Toscana, nor is one apparent in **[96]** (page 108). One could easily have been added for Bosisio if there were concerns over the accuracy of the original gauge or abnormal oil pressure after his engine rebuild.

Asked if he had made changes in the arrangement of the instruments, Bosisio replied negatively in an understatement:

> *No, that was from Ferrari. Ferrari and Touring did that. The instruments were left the way the car was born. We only redid the front. We built the roof and the back that was slightly different. Instead the instruments and the steering wheel and what was working of the car was left the way it was made by Ferrari Touring. We left it original.*

Since they were giving the car a new look, it could have been that Zagato tried a new layout and then rejected it while work was underway. The same instrument configuration, including the new instrument but not the extra oil gauge, was used again by Zagato the next year on the A6G 2000 Maserati Spyder prototype, although it was a mirror image for a left-hand drive car.[123]

The new instrument was a tachimedion, a recent invention by Fratelli Borletti that could mechanically compute the car's average speed over distance and time. Such information would be useful in long distance races. Having that instrument and the speedometer/odometer in front of the passenger, who can function as a navigator, was a practical arrangement, but it was even more practical having them together in that the tachimedion was driven by a cable attached to the speedometer. The device was described in an August 1952 *Popular Science* magazine article **[127]**. *Auto Italiana* also printed a lengthy article in the February 14, 1953, issue which followed the work on 0046M but indicates its avant garde appeal.[124] Unfortunately, Hans

[123] Walter Bäumer, *Maserati A6G 2000 Zagato*, p. 62. Chassis number 2101.

[124] "Inaudi a bordo di ogni automobile (Il 'Tachimedion' VEGLIA)," pp. 39-40.

Tanner observed that "these instruments had an annoying habit of going wrong after a week, and from then on never seemed to function at all."[125]

On the driver's door, inside, Zagato added a novel feature: A knee brace was welded to the tubular frame of the door, padded and covered with what appears to be black leather to match the dash. This provided a rest for the driver's right knee in long distance races and a way for the driver, in Bosisio's words, "to plant it while turning" **[128-129]** and **[141]** (see page 141). To further describe the interior, here follows my questioning of Bosisio through Mary Virginia Reese in 1989, newly transcribed from the recording and translated in 2018 by Giovanni Zimotti, a University of Alabama doctoral student from Italy who has assisted me. Direct quotes from Bosisio are in italics:

MVR: Was it red when you raced it?

Yes, it was all red. Red, the Ferrari color, like the ones that are currently racing in Formula 1. A dark shade of red, a strong red.

MVR: Was it red inside?

No, it was not red inside. It was of a natural color. A light color.

MVR: Was there a rug?

Yes, there was a car mat. Some sort of carpet, but not really carpet. A rug a little bit rougher, but also lighter. Then the gearshift knob was ivory.

MVR: What about on the doors, was there something on them?

I do not recall there was anything special. Maybe they were painted with a paint that was sound-deadening.

[125] Hans Tanner, *Ferrari Owner's Handbook*, p. 11.

TO START computing odometer, flip down lever (left). Knob on its pivot returns average-speed pointer to "zero position." Stem underneath winds clockwork. Fittings (photo below) are installed in half-hour.

New Car "Clock" Averages Speed

IF YOU drove your car at a steady 63 miles an hour for 28 minutes, 49 miles an hour for 52 minutes, and then took 33 minutes out for lunch, what do you estimate your average speed—including the stop—would be?

It's probably less than you think. A new instrument for your panel, a computing clock-odometer, will tell the average at a glance—slightly less than 38 miles per hour.

The Tachimedion will average your speed over hundreds of miles. If you want an average on driving time only, it will give that.

It mechanically divides distance traveled by time elapsed. A series of cams and gears comes up with the answer. The time is fed into it by an integral spring-actuated clock. It receives the distance factor from the car's odometer cable.

Standard equipment on Italian-made Alfa Romeo and Lancia cars, the instrument sells in the U.S. for about $90. It is marketed here by Imperial Motor Products, 570 Fifth Ave., NYC 19.

[127] 1952 *Popular Science* article describing the tachimedion. *Popular Science*

[128-129] Knee brace added by Zagato on the driver's door. *Author*

A paint that was a little bit wrinkled. Light [weight], *but that was holding back the noise.*

MVR: Were the seats covered with fabric?

Yes, they were covered with fabric.

MVR: What color?

I think a light color, because outside it was dark red and inside a very light color fabric.

The "fabric" of "light color" Bosisio described indicates that the seats were recovered. Since a headliner was put in the new top and it was also a light-colored material, it was probably fabric also. There will be mention of fabric-covered seats in later years, as well. The seat material could well have been corduroy, commonly used to lessen sliding about in an era without seatbelts, and the color could well have been tan, a popular choice.

By no means was the car's interior spartan. The floor mats which Bosisio mentioned were fitted and bordered, and coverage extended behind the seats to the shelf above the fuel tank and to fitted pieces over the inner rear fender wells, the remains of which can be seen later in **[246]** (see page 229). There were delicate aluminum trim pieces painted flat black and shaped to border the headliner around the windshield, over the doors, around the rear quarter windows and around the back window. The insides of the B-pillars were covered with polished aluminum trim pieces and there was a polished aluminum bow securing the headliner to the frame ahead of the back window. For its day, 0046M had become a comfortable and stylish little coupe.

All of the construction, modification and trim work took a great deal of planning, care and skill to make everything fit together perfectly and appear as though it had always been that way. It is no wonder I never thought of the possibility that the car could have been a *barchetta*.

Estimating what Zagato charged for the transformation would be pure conjecture. As a comparison, a 1956 issue of *Road & Track* stated that Carrozzeria Scaglietti charged about $3,000 for an entire custom body.[126] In 2024 dollars, that would be about $34,800, but Bosisio's cost for the partial work would have been far less. Post war, skilled labor was cheap in Italy.

The *licenza circolazione* of MI 206517 was issued on January 20 and the first race for 0046M in its new shape was a hillclimb in Verona on March 29, the Corsa in Salita sulle Torricelle or the "Torricelle Uphill Race."[127] Contestants began their run in the valley below the Santa Giuliana ridge north of Verona up to the *torricelle* above, which were "small towers" built as defensive gun emplacements by the Austrians in 1837. The distance was 4.7 kilometers, rising 200 meters over 27 curves.[128] Bosisio placed 4th in the over 1100 Sports class with a time of 00:03:23.6 and only three seconds behind the overall winner for 7th.

On the back of the speedometer, I found a tag that read "Ruggeri a Nizzi 21 Apr 1953," and when I asked Bosisio about that he said,

I had it fixed. It is the name of the company that fixed it. It is a company that specializes in odometers.[129]

That date indicates the car was being readied for the Mille Miglia on April 25, and it could have been that he wanted to match the speedometer to the ratio of the differential for an accurate measure of speed and distance traveled.

Zagato's finished coachwork is shown in **[130]** and **[131]**, first as the crowd admires the "new" car in Brescia and then as Bosisio puts 0046M in the lineup for the XX Mille Miglia bearing the 5:20 AM start time number. Standing beside the open passenger door is his teammate, fellow Milanese Pier Luigi Ercole, three years younger but a veteran co-driver of three previous Mille Miglia races.[130] Giletti was five cars

---

126 Dregni, p. 64.

127 Marchianò misidentified the V Coppa della Toscana as the first race for the new *berlinetta*. *Ferrari by Zagato*, p. 59.

128 Danilo Castellarin, *Correndo nel Tempo: Storia del motorismo veronese nel Novecento*, p. 15.

129 It was the exclusive representative of Smiths Instruments in Italy at Via Andrea Doria 32, Milan.

130 If not then, Ercole was later employed by Bosisio's company in clerical work and they remained adventuring companions, much later surviving an airplane crash together.

back, starting at 5:25 AM in a Maserati A6G 2000S, and since he must have noticed 0046M, one has to wonder what he thought about his car from last year.

The new registration number is affixed to the ventilated shield over the exposed bottom of the radiator. There are now safety buckles on each side at the back of the hood[131] and ahead of each door, above the waistline, is a decal of the Scuderia Sant Ambroeus insignia at the time **[132].** Clearly seen in the cabin are the light-colored headliner and the polished, curved aluminum trim that attached it to the frame ahead of the back window.

Figure **[133]** is a beautiful photograph of 0046M on the starting ramp near dawn on April 26. Behind them in the little cabin they have draped a screen to block the glare of headlights coming through the rear window. Jess Pourret, my original source for this image by Sorlini, commented on it in a 1990 letter, saying,

[131] Adding leather straps and buckles to rear-opening hoods was a common practice to prevent the guillotine effect they could have in a crash.

[132] Emblem of Scuderia Sant Ambroeus. *Zagato: Seventy Years in the Fast Lane*

[130] XX Mille Miglia, 1953 – 0046M with its new bodywork. *Luigi Bosisio*

[131] XX Mille Miglia – Bosisio puts 0046M in the lineup for the start with Pier Luigi Ercole at the passenger door. *Revs Institute*

XX
LLE MIGLIA
-26 April
195
SUPERCORTEMAGGIORE
fente benzina italiana
206517•MI
IV

The influence of the Marzotto brothers can be spotted because the car is shod with Michelin 'X' tires, which allowed the brothers superiority over the competitors shod with the poor Pirelli 'Stelvio' but Ferrari was then under contract with Pirelli and couldn't use the Michelin X's for the factory cars. They made a monumental difference, in particular with wet roads.

The streetlights are still on but the sky is growing lighter with the approaching sunrise and an official waits for the second hand of his watch to close on 5:20 AM, when at his call the flagman will signal the car to be away in the coming dawn.

Figure **[134]** shows them on the road but they did not get very far. They made it the 160 km to Padova, at least, as the photograph taken there attests **[135]**, but not to the first checkpoint in Ravenna because at some point, as Bosisio said, "I broke a drive shaft. It is one of the two shafts that transmit power to the drive wheels." By which he meant one of the rear axles broke. His original time card for the Mille Miglia **[136]** would have been stamped at each checkpoint had he reached them. Giletti, on the other hand, had a fine race and finished 6th overall.

On May 17, axle repaired, Bosisio was on the island of Sardinia to enter another hillclimb, the Scala di Giocca-Osilo. Beginning at an elevation of about 140 meters a few kilometers to the southwest of Sassari at an area known as Scala di Giocca ("Giocca staircase"), the road ascended as the name describes in a long, climbing series of a half-dozen hairpin switchbacks about 200 meters up the face of a cliff. From there the road set off to the village of Osilo, about 10 km away over mountainous terrain at an elevation of 672 meters. Bosisio placed 2nd in the 1100-2000 cc class with a time of 00:10:10.9. Placing first in his class was the ever-present Diego Capelli with a time of 00:09:46.7 in his uncle Ovidio's bespoke 8V Zagato roadster.

A layover of a week in Sardinia could have allowed practice on the road between Sassari and Cagliari on the south coast

[133] XX Mille Miglia – Bosisio and Ercole on the starting ramp waiting for 5:20 AM. *Giorgio Nada Editore Archive*

[134] XX Mille Miglia – On the road. *Giorgio Nada Editore Archive*

[135] XX Mille Miglia – Passing through Padova. *Primavera Bresciana XX e XXI Coppa F. Mazzotti Mille Miglia*

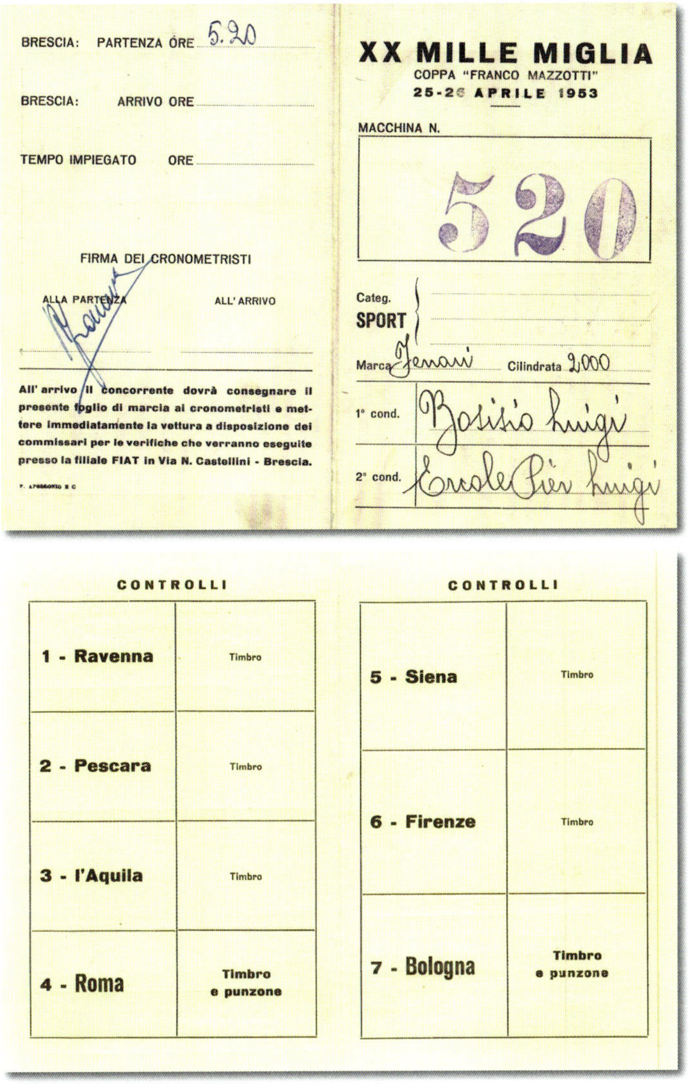

BRESCIA: PARTENZA ORE 5.20

BRESCIA: ARRIVO ORE

TEMPO IMPIEGATO ORE

FIRMA DEI CRONOMETRISTI

ALLA PARTENZA ALL' ARRIVO

All' arrivo il concorrente dovrà consegnare il presente foglio di marcia ai cronometristi e mettere immediatamente la vettura a disposizione dei commissari per le verifiche che verranno eseguite presso la filiale FIAT in Via N. Castellini - Brescia.

XX MILLE MIGLIA
COPPA "FRANCO MAZZOTTI"
25-26 APRILE 1953

MACCHINA N.
520

Categ. SPORT

Marca Ferrari Cilindrata 2000

1° cond. Bosisio Luigi

2° cond. Ercole Pier Luigi

CONTROLLI

1 - Ravenna	Timbro
2 - Pescara	Timbro
3 - l'Aquila	Timbro
4 - Roma	Timbro e punzone

CONTROLLI

5 - Siena	Timbro
6 - Firenze	Timbro
7 - Bologna	Timbro e punzone

[136] Bosisio's time card – unstamped because he never reached the first control point. *Luigi Bosisio*

where on May 24 the III Trofeo Sardo race began. Bosisio was joined by seven other members from Scuderia Sant Ambroeus, and since the entry list says he had a co-driver,[132] probably it was Pier Luigi Ercole with him again. As with the year before, the race was in two legs, the first from Cagliari to Sassari, where there was a rest stop and staging for the return leg to Cagliari. The race began with a staggered 10:00 AM start with Bosisio flagged off at 10:09, but 0046M was simply numbered 20. Teammate Elio Zagato recalled an amusing incident on his way to Sassari:

[132] Alfano, p. 77.

> *... we reached a railroad crossing and slowed down, seeing that a train was coming. To our astonishment many locals who were watching the race gestured for us to continue fast and, looking more carefully, we realized that the bars of the railroad crossing had been mounted* [across the track] *in order to give way to the road vehicles. We had the great satisfaction of having taken precedence over a train.*[133]

If only the train tracks had been made so accommodating for Bertone in the 1950 Coppa della Toscana.

Bosisio arrived in Sassari with a time of 01:35:00, good enough for second in his 2000 Sport class, but Diego Capelli had again bested him with a time of 01:33:24. Both probably benefitted from the early retirement of Giletti in a Maserati A6GCS.

[133] Zagato, p. 53.

The rest of the *scuderia* had acquitted themselves well at midway with their five Alfa Romeo 1900 TIs in the 2000 Touring class. Arriving in 1st place in class was Piero Carini, a professional driver of considerable skill who was also employed that season by Scuderia Ferrari and who had barely outpaced Bosisio by 14 seconds with a time of 01:34:46. The other Alfas followed in 2nd, 4th, 5th and 6th in class, and their lone Fiat 1400 was 3rd in 1500 Touring. A photograph from Bosisio's personal album **[137]** captures the group at the staging area along via Serra Secca.[134] The festive Bosisio, wearing a tie in the center rear, is lifted up by hugging Elio Zagato with his left arm and Franco Martinengo, Zagato's co-driver, with his right. Both were tall men. The woman on the right is probably Olga della Beffa, frequent co-driver and I believe the sister of Alberto della Beffa,[135] who is squatting in the center, and on the far left is Piero Carini. Peeking between Elio and Olga is 0046M. In almost exactly where the Ferrari had sat the year before **[138]**, Bosisio comments at the back while Olga and Elio stand behind the car. Carini is examining the interior of 0046M while his Alfa Romeo number 42 sits behind, although arriving ahead of Bosisio. Carini was probably impressed by the close race 0046M had given him, and his interest in the car is important to note for later developments.

[134] Judging from the partial license plate, the Ferrari behind them is possibly 0022M, which was eventually owned by Cooper Weeks and brought to the Fortieth Anniversary Reunion of the *barchettas*.

[135] A sportsman and founding member of Scuderia Sant Ambroeus, Alberto della Beffa placed 10th for Italy in both two and four man bobsled at the 1952 Winter Olympics. He and Olga placed 31st in the Alfa earlier in the 1953 Mille Miglia, two places behind Zagato and Martinengo, another founding member.

[137] III Trofeo Sardo, 1953 – The Scuderia Sant Ambroeus team at Sassari. Standing left to right: Piero Carini, Franco Martinengo, Bosisio, Elio Zagato, Olga della Beffa; below: Alberto della Beffa. *Luigi Bosisio*

[138] III Trofeo Sardo – The team examines 0046M with Carini looking inside. *Luigi Bosisio*

At 3:00 PM, the cars began to depart at intervals on the second leg back to Cagliari. In the 2000 Sport class, Diego Capelli maintained his lead, but Bosisio had narrowed the gap to one minute at the 140 km checkpoint. As the race progressed to the south, rain had been falling there incessantly. At 700 meters before Sanluri, only 45 km from the finish, Capelli lost control on a high speed curve in the wet, left the road, and was killed in the crash. One wonders what Bosisio's thoughts were as he passed the site of the fatal accident and pressed on in possession of 1st place in class, which he held to the finish with a total time of 03:16:35.3. Carini was unchallenged for 1st place in his class, followed in 3rd, 4th, 5th, and 6th by the Alfas of his Scuderia Sant Ambroeus teammates. Bosisio's 1st place in the 2000 Sport class won a prize of 160,000 lire.[136] The prize was worth about $260 then and about $3,070 in 2024 dollars.

Unlike Giletti's race there the year before, this time there were contenders with engines well over 2 liters and the results bore evidence of that. In overall standings, Carini finished 5th, and the other Sant Ambroeus Alfas were 7th, 8th, 9th, and 10th, while Bosisio was 11th. Indicative of Ferrari's state of affairs in the 2000 Sport class, Bosisio averaged 128.9 km/hr while Carini averaged 131.2 km/hr in the 2000 Touring class Alfa sedan.

[136] Alfano, p. 90.

The following weekend, May 31, Bosisio was back on the mainland in Livorno for the V Coppa della Toscana. His start time number 917 is being painted on the car before the race as Bosisio socializes wearing a light-colored pullover **[139]**. Bosisio, now wearing a sweater, brings 0046M to the start, then he and his co-driver, probably Ercole again, don their helmets, board the car and depart with a roar that causes one onlooker to plug his ears **[140-144]**. Figures **[141]** and **[144]** are the photographs signed by Elio Zagato and sent to me through Michele Marchianò.

Wet conditions persisted in the race as 0046M goes through Livorno, along the seaside road heading south, passing a billboard advertisement for a butane gas company, going through countryside and a control point until the rain finally stops **[145-151]**. While his recall of statistics was inaccurate after 35 years, Bosisio had an indelible memory of the difficulty he faced:

> *It was a very demanding race. The Coppa della Toscana was 730* [actually 633] *kilometers long... and in that race, it rained like if it was the Great Flood. 121* [200] *cars started the race and maybe 20 or 40* [109] *arrived at the finish line, everyone else was involved in a crash.*

In the end, Bosisio finished 3rd in class and 28th overall with a time of 06:09:53.

On June 28 at the Chiusaforte-Sella Nevea hillclimb, Bosisio winds his way up mountain roads in **[152-153]**. Chiusaforte is located at an elevation of 391 meters in the province of Udine, the far northeastern corner of Italy where the borders of Austria and Slovenia meet, and the 13 kilometer road to the village of Sella Nevea rises to an elevation of 1,161 meters. In **[154]** he charges across the finish line, car 92 taking the

[139] V Coppa Della Toscana, 1953 – Bosisio socializes while his race number is painted on. *Luigi Bosisio*

[140] V Coppa Della Toscana – Bosisio and Ercole push the Ferrari up to the start line. *Luigi Bosisio*

[141] V Coppa Della Toscana – Bosisio settles in the car for the race as Ercole gets in. Signed by Elio Zagato. *Luigi Bosisio*

[142] V Coppa Della Toscana – A crowd around the start line; flag man at the right. *Luigi Bosisio*

[143] V Coppa Della Toscana – Helmets on, Bosisio and Ercole are ready. *Luigi Bosisio*

[144] V Coppa Della Toscana – With a roar that causes one onlooker to cover his ears, they are off. Signed by Elio Zagato. *Luigi Bosisio*

[145-146] V Coppa Della Toscana – Passing through Livorno. *Luigi Bosisio*

checkered flag with a time of 00:14:24, good enough for 6th in the over 1100 Sports class and 20th overall, compared to the class winning time of 00:13:17.6 by Gerino Gerini in a 212 Ferrari, 0237EU.

From that point on, I have yet to find a record of Bosisio racing 0046M again in 1953, and although he twice mentioned having raced 0046M at Monza, I have found no proof of that. He was still racing, though. He was a co-driver at the 10 Hours of Messina on July 26 in a Fiat 8V, and again a co-driver in the prototype Fiat 8V Zagato at the 12 Hours of Pescara on August 15.

[147] V Coppa Della Toscana – South of Livorno by the Ligurian Sea. *Luigi Bosisio, photographed by Foto Florenza*

[148] V Coppa Della Toscana – Wet roads in the countryside. *Luigi Bosisio, photographed by Foto Florenza*

[149-150] V Coppa Della Toscana – Stopping at a control point. *Luigi Bosisio*

[151] V Coppa Della Toscana – Passing through towns and villages. *Luigi Bosisio*

[152-153] Chiusaforte-Sella Nevea hillclimb, 1953 – Bosisio negotiates the rocky, narrow road up the mountains. *Luigi Bosisio*

In retrospect, Bosisio was quite pleased with 0046M and the Zagato revision. In his words, "I raced so many times with that car. It was very enjoyable. That car gave me great satisfaction." Of its best attributes, he named the aerodynamics – "it was faster" – and when asked about its disadvantages, he said, "Maybe the brakes. It didn't brake so well." And he added, "I really enjoy talking about this topic and I get emotional. Also, if [you] bring it to Italy, we could race an historic race together."

Like Giletti before him, Bosisio's racing ability with 0046M caught the eye of Maserati and he too was taken on as a driver the following year with modest success. Again like Giletti, in 1955 he too retired from racing to concentrate on business.

Michele Marchianò summarized the significance of 0046M well in our telephone conversation on November 17, 1988:

> *It has much history and in fact its beauty is historic. Few cars can claim the history of both Touring and Zagato. It is the fusion of two beautiful cars in one machine. Bertone owned it. Touring designed it. Zagato re-did it. Thus, the world's three greatest carrozzeria put their hands on this car. Also this car won big races and ran in the Mille Miglia. The style was important. It is, in short, a historical document.*

[154] Chiusaforte-Sella Nevea hillclimb – Crossing the finish line.
Luigi Bosisio

AUTOMOBILE CLUB DI

Chapter 7

1953-1955: MARTIGNONI AND THE EXPEDITION TO AFRICA

0046M

The records of the Varese *Ufficio del Pubblico Registro Automobilistico* show that the sale of 0046M from Bosisio to Bruno Martignoni of Varese was recorded on December 9, 1953 (Appendix 5). Sadly, Martignoni died in 1986 before I began reaching into Italy to contact the early owners, so I have no direct testimony from him about his time with the Ferrari. I was fortunate, however, to have been alerted in 1991 to the publication of a book about his racing career, *Martignoni l'africano* by Robi Marzoli, and immediately purchased a copy. With that and assistance from other sources, I believe I understand the circumstances surrounding his purchase and use of 0046M.

Unlike Bertone, Giletti and Bosisio, who were all young men when they purchased 0046M, Martignoni was 40. His father, Enrico, died when he was four years old and his mother, Maria, remarried the widower Umberto Isella in 1919 when he was six. That year, Isella and his brothers formed a transport and moving company, first with horses, then trucks and then began to diversify into other businesses.[137] The early exposure to his family's commercial vehicles may have kindled Martignoni's attraction to automobiles which eventually led to his ownership of a Lancia sales and repair facility in central Varese, along with other interests.[138]

Marzoli portrays Martignoni as being from a well-to-do family and spending much of his time as a youth in the fashionable cafes, eventually becoming fascinated by the adventure of racing and, in spite of his parents' entreaties, taking to the sport.[139] His inclination to go fast was evident in 1939, when at 26, he sped into a collision which injured himself, his passenger and six others.[140] According to the registration of the Ferrari sale, in 1953 he lived in the impressive home Umberto Isella built on the street where the family business began at 42 Via Cavour in Varese, a four-floor building that still shows its contemporary elegance, although now converted to shops and offices **[155]**. By

[155] 42 Via Cavour, Varese, today – The Isella residence where Martignoni lived with his mother and stepfather. *Google*

then, he had been racing for seven years, beginning in 1946 with Fiat 500-based racers and moving up in displacement and success from Fiats to Lancia Aprilia to Siata and finally to Alfa Romeo 1900, winning frequent firsts in class. A handsome man who dressed well and seems to have been a popular and charismatic bon vivant, Martignoni was well known in racing circles. He was a mentor to Elio Zagato as the coachbuilder's young son began his racing career, and in his autobiography, Zagato recalled how Martignoni had introduced him to a benefactor in 1947:

I admired Bruno Martignoni, of whom I became a friend.

[137] www.museoweb.it/imprese/isella/#tabs-10, accessed June 23, 2022, link now defunct.
[138] Private email from Alberto Bianchi, Martignoni's nephew, July 19, 2022.
[139] *Martignoni l'africano*, pp. 11-12.
[140] *Corriere della Sera*, July 10, 1939.

> *He was a good driver and assiduous participant in the endurance races in Italy and abroad. Since I had started to get interested in racing and race cars, Bruno wanted to introduce me to Giorgio Giusti who, in addition to racing, was involved in racing engines and the Race Club 19 in Turin. Bruno said to me: "Let's go to his house, see if we can get you acquainted. Anyway, you'll get to drink the best cognac I know." Well! Giorgio became my great friend and I visited the family and his house for a long time.*[141]

To explain what happened during Martignoni's tenure, it is necessary to put other events in perspective, since 0046M played a small part on a much larger stage. For 1953, the *Fédération Internationale de l'Automobile* (FIA) had instituted a World Sports Car Championship, which consisted of seven races contested by both private and factory teams, and since it was indeed a world championship, the races were held worldwide. Ferrari skipped the first event in March because Enzo "had judged that the winter expedition to Florida for the Sebring 12 Hours was too expensive for a company that remained small, and in which all expenses must be carefully weighed."[142] Indeed, all of the manufacturers skipped various races for budgetary reasons.

By the 5th round at the Nürburgring 1,000 Kilometers on August 30, Ferrari had fallen behind Jaguar in the points standings, prompting a push with three factory cars pairing Alberto Ascari and Nino Farina, Luigi Villoresi and the lone non-Italian, Mike Hawthorn, plus the "B" team drivers of Umberto Maglioli and Piero Carini. The cars were three 4.5-liter 375 MMs. Also at the race, and no doubt socializing with his Italian compatriots, was Martignoni, who had entered an Alfa Romeo 1900 with a co-driver, Sergio Mantovani. It seems to have been Martignoni's first race with the Alfa Romeo 1900 and it became his choice for racing from then on, with one exception. The 375 MM of Ascari and Farina won the race and brought Ferrari into a points tie with Jaguar. Martignoni finished 24th, but more noteworthy was his conjunction with the Ferrari factory teams.

Jeopardizing Ferrari's outcome in the Championship was Enzo's declaration after Nürburgring that he was withdrawing from racing, a bluff that he would again resort to in the future to achieve his ends. Ever conscious of finances, this time his complaint was that the FIA had approved eliminating appearance and prize money from all events.[143]

The next race of the series was the Tourist Trophy, held far away in Ireland, and as luck would have it, the Jaguar team had only a third place finish there, the points for which were discarded under the rules since only the four best results were counted.

As the 1953 season wound down, then, Ferrari was still tied with Jaguar for the World Sports Car Championship, and the 7th and concluding Mexican Road Race, or Carrera Panamericana, from November 19 to November 23 would be the deciding event. In addition to Enzo's vow, the cost of intercontinental logistics was a factor and Scuderia Ferrari did not enter the race. Similarly affected by the cost, Jaguar also declined to participate with a factory team. For both, however, private teams came through to be their representatives. As they had the previous year, Ferrari relied on Scuderia Guastalla for the race and provided them with the factory competition cars – four 375 MMs – and one Ferrari driver, Maglioli.[144] A Ferrari placed 4th, enough to clinch the Championship, and afterward the entourage remained and moved on to two other races in South America.

By the end of 1953, Ferrari was wrapping up another fabulously successful year of racing. They had won the Grand Prix World Championship again with Alberto Ascari, while his teammate Giuseppe "Nino" Farina was third. Factory and customer cars were winning races all over Europe and they had won the World Sports Car Championship. Across the Atlantic, Ferraris were beginning to make their mark in America and they had seen success in the Carrera Panamericana and Argentine Formula Libre races.

On December 12, the annual company luncheon was held in Modena to celebrate the victories and championships of

[141] Zagato, p. 23.
[142] dal Monte, p. 466.
[143] Ibid., p. 465.
[144] *Ferrari Automobili 1947-1953*, p. 353.

the year and, in a gesture that was more self-serving than magnanimous, Enzo announced that Ferrari would return to racing. Without a clear understanding of their future, the Ferrari drivers, Ascari and Villoresi in particular, had grown restive, and the announcement was meant to reassure them of employment.[145] It was also simply an opportune way to justify plans that were already well underway for 1953's final race, the 12 Heures de Casablanca eight days later on December 20.

Across the Mediterranean in Africa, racing was beginning to reawaken after World War II. In the 1930s, Enzo's Scuderia Ferrari had raced Alfa Romeos in Libya, Morocco and Tunisia, but postwar Ferrari efforts on that continent seem to have been limited to private entries. For example, Frenchman Jean Lucas managed one win, in the Rallye du Maroc (Morocco) in 1951 and Pierre Boncompagni (known as Pagnibon) had a win at the 3 Heures d'Alger in May 1953.

The new Scuderia Ferrari was now ready to resume racing in Africa but the cars loaned for the Carrera Panamericana race had not yet returned and the stable was empty. Nonetheless, the "A" team drivers were still available and wanting action: Ascari, Farina, Clemente Biondetti and Villoresi who were a group of proven race winners. So, customers who had bought the latest racing machinery were enlisted and paired with the factory drivers, which was the opposite Ferrari had done for the Carrera Panamericana but equally cost effective as a joint effort with privateers. Ascari was to be with Casimiro de Oliveira in his 375 MM, Farina with Piero Scotti in his 375 MM, Biondetti with Luigi Piotti in his 250 MM and Villoresi with François Picard in his car, the first 500 Mondial, sent for its maiden race as Ferrari's new four-cylinder contender in the 2-liter class.[146] It was an all-out invasion of North Africa for what was essentially an unimportant race, but it served Enzo's purpose with his drivers and made a statement.

Martignoni was very familiar with North Africa, having placed 1st in class at the Rallye du Maroc in a Lancia Aprilia in 1951, and in 1952 he had finished 5th overall and 1st in class in a Fiat 1100 at the previous 12 Heures de Casablanca. His real notoriety, though, came from his overall victory for the Fiat factory in the II Rallye Alger-Le Cap, a rally lasting 25 days, beginning at Algiers on the Mediterranean on December 28, 1952 and ending at Cape Town on the southern tip of Africa on January 21, 1953. Leading a three-man team in a factory-prepared Fiat 1900, Martignoni traversed the length of the continent and traveled 16,000 km through deserts, plains and tropical forests, usually on unimproved roads. Newspapers reported their progress and upon returning to Italy, *La Gazzetta dello Sport* dubbed him *Martignoni l'africano* in the headline of a feature article and interview.[147] In April, Fiat exhibited their car, named "Kon Tiki," at the Turin auto show, so it was a celebrated victory, bringing Martignoni fame.

Somehow, Martignoni became attached to the Ferrari expedition. Like Ascari and Farina, he was an old hand and already knew them, at the least by having raced against them at the Nürburgring four months before. Farina in particular would have felt an affinity for Martignoni. Farina has been described as, "Well-mannered, charming and gracious on most occasions, [but] he could also be arrogant and aloof. He was accused of being unsentimental and a snob who disapproved of those members of his profession who did not have the right social pedigree."[148] Indeed, his nickname was "the Gentleman of Turin" and he likely saw Martignoni as a worthy companion.

Martignoni had settled on the Alfa Romeo 1900 when it came out, but for the opportunity to tag along with Scuderia Ferrari, he had to have a Ferrari to race himself. Possibly to cement his relationship with the Ferrari team, and certainly to make his effort competitive, he took – or was given – Piero Carini as his co-driver for the 12-hour race. Carini was introduced in the previous chapter and had been a yeoman driver for all of the major racing teams. With Maglioli off to Mexico, he was the remaining "B" team driver of Scuderia

[145] dal Monte, p.475.

[146] Respectively, they were: 0366AM, 0360AM, 0276MM, and the 500 Mondial might have been 0302TF, although there is some question about its serial number.

[147] February 26, 1953.

[148] *www.motorsportmagazine.com/database/drivers/giuseppe-farina,* accessed June 15, 2022, link now defunct.

Ferrari, so he was an asset, and he knew where there was a used Ferrari available that still had some racing life left, particularly one with a large capacity fuel tank ideal for an endurance race.

When I asked Bosisio about his sale of 0046M, he insisted that he sold it to Piero Carini. To quote him: "He [Giletti] sold it to me and then I sold it directly to Carini. So, I do not think that it went through Martignoni."

Undoubtedly, Bosisio knew Carini, who at 32 was just a few years older. They both had raced under the Scuderia Sant Ambroeus banner at the Trofeo Sardo earlier in the year and had a close race with each other to Sassari, Carini barely winning out. Carini's curiosity about 0046M and its ability to keep up with him was evident in **[138]** (see page 139), and it should be noted that he had raced 166 MM 0004M previously, so he must have recognized that this was not the ordinary 2-liter Ferrari. With his work for Ferrari at a standstill, Carini had capped his European season for the year on November 1 with a 2nd place at Monza in Scuderia Sant Ambroeus' culminating Coppa Sant Ambroeus race with his Alfa from the Trofeo Sardo. There he might have heard that Bosisio was in the market to sell 0046M.

It is worth considering that since Elio Zagato knew Bosisio, Martignoni and Carini, he too could have played a part in the car's changing hands.

Surely it is possible, if not likely, that Carini served as an intermediary for Martignoni, and that might be why Bosisio said 0046M went to him. The Public Register says the car was actually sold on December 9, 1953, and on December 10 the sale was recorded in Busto Arsizio, a town north of Milan. Then on December 11, the record of the transaction was transferred, as was customary, to the central *Ufficio del Pubblico Registro Automobilistico* for the province in Varese. It was noted as a private sale with no figure listed from Bosisio to Bruno Martignoni, son of Enrico. There, Martignoni paid the usual 10,557 lire tax on the sale ($17 in the 1953 equivalent or $200 in 2024). No *licenza circolazione* was issued since the race was looming in nine days on the 20th and the Ferrari must have been whisked away to Casablanca posthaste.

Equipped with his charm, his recent celebrity, his racing experience, his knowledge of the course and country, and now a Ferrari, Martignoni appears to have become part of Ferrari's expeditionary group in some capacity, whether it was official or unofficial. This is not to say that he was representing the factory; the others were entered by Scuderia Ferrari while his was a private entry. He might have been seen as a backup in the 2-liter class to the unproven 500 Mondial and that may explain why his name was listed as second owner of 0046M after Bertone in the information Ferrari sent to me in 1966. Quite likely, the Scuderia Ferrari mechanics prepared the car for the race and the factory records reflected Martignoni as another beneficiary of Ferrari's largesse.

In the spring of 1996, I wrote a letter to Martignoni's Varese address at 42 Via Cavour, asking if anyone living there could offer information. To my great delight, Umberto Rusconi, a nephew of Martignoni, replied in July. He said in part,

> *I remember (but it is only a remembrance of my youth) that I came from Milano to Varese in the Ferrari the very day that my uncle bought it. I remember too that, in that very year (1953), I met Alberto Ascari who came to Varese to arrange his participation and my uncle's with Villoresi, Farina, Scotti and Carini to the 12 Heures de Casablanca.*

Ascari's granddaughter, Veronica Ascari, has related the World Champion's opinion of Martignoni, saying Bruno Martignoni was among those "esteemed by him."[149] Enough so that he would travel to meet with him, which confirms that Martignoni was closely involved with the Ferrari effort and was being consulted.

The photographs **[156]** and **[157]** are reproduced from the book *Martignoni l'africano* and portray him as one of the crew in Casablanca.[150] In the first, from the newspaper *Le Petit Morocain*, Farina (left) and Martignoni relax together. In the second, the caption reads, "On the back of this photograph Martignoni noted in handwriting, 'Biondetti-

---

[149] *La Prealpina*, January 7, 2017.
[150] pp. 51, 54.

Villoresi-Farina carefully count the Moroccan Francs they have received for the engagement.'" Farina is on the right in the checkered jacket, Martignoni is beside him on his right, and both are handling the money. Martignoni appears to be handing money to Villoresi, while Biondetti, across from Farina, waits for his share.

[156] Nino Farina (left) and Martignoni in conversation at Casablanca. *Le Petit Morocain*

[157] Counting out the Casablanca appearance money. On the left, Biondetti and Villoresi; on the right, Farina and Martignoni. *Bruno Martignoni*

Casablanca was likely a favorite city for a voluptuary like Martignoni who may have shown his teammates what it had to offer. Duncan Hamilton, a renowned race driver of the era, spent time there the following year and in his book *Touch Wood* recounted his amusing experience at the Hotel Sphinx, which turned out to be a brothel, but on the other hand he said there was danger to be had from thieves, highwaymen and political terrorists.[151]

[151] pp. 152-156.

Unlike most circuits and road courses in Italy and Sicily, Ain Diab, the Casablanca circuit in the suburb of that name, was flat. Located on the coastal plain and bordered on its northwest side by the Atlantic Ocean, the elevation was only 15 meters and the course consisted of 3.26 kilometers of public roads, basically in the shape of a rectangle, so it was very fast and considered dangerous. In practice, de Oliveira crashed his 375 MM and it was lost for the race, so Picard graciously gave Ascari his place in the 500 Mondial, leaving him a spectator to his own car.[152]

As with many endurance races of the day, the drivers began the 12 hours with a Le Mans start, running across the Boulevard de l'Ocean Atlantique to the 30 cars **[158-160]**. As the cars begin to pull away, 0046M can barely be seen, the 6th car still parked on the left, behind the number 19 Veritas of Oswald Karch. By **[160]**, 0046M is safely on track but balked behind the entering number 3 250 MM of Ferrari teammates Luigi Piotti and Clemente Biondetti. Later, 0046M (number 22) is leading the Fiat 8V of Mario Dustaritz and Yves Dupont and about to overtake the smaller class Deutsch-Bonnet of J. Ferreira da Silva and Corte Real (?) Pereira **[161]**. The scene in **[162]** appears to be a pit stop, but since Martignoni is not wearing a helmet, the photograph is probably from late at night after the race. The headliner over the passenger seat has come loose, perhaps from the wind buffeting through the cabin.

From the beginning, Farina and Scotti in the 375 MM took the lead of the race and never relinquished it. In the 2-liter class, the new 500 Mondial of Ascari and Villoresi held that lead through most of the race. By the 4th hour, the Gordini of Roger Loyer and Charles Rinen ran 2nd with the OSCA of Jean-Louis Armengaud and Élie Marcel Bayol 3rd and Martignoni and Carini trailing. By the 8th hour, Martignoni and Carini had moved up to 3rd, ahead of the OSCA. At the 10th hour, the Mondial was forced to make a pit stop, allowing the Gordini to take over 1st place, but Ascari gave chase, caught up, and pressured the Gordini into a mistake and out of the race in the 11th hour, regaining 1st. Meanwhile, the OSCA overtook Martignoni and Carini to take 2nd place, relegating them to 3rd in class, and that was the order at the end of the 12th hour and the race.[153] Only 17 cars finished, fewer than half the field.

Due to a number of mechanical failures and fuel issues, four of the over 2-liter class cars dropped out, which benefitted Martignoni and Carini who finished 7th overall. In serious contention with the big cars, though, was the 500 Mondial,

[152] Francesco Scaletti, *Ferrari I Quattro Cilindri*, p. 161.

[153] *Auto Italiana*, December 21-31, 1953, pp. 23-25.

[158] 12 Heures de Casablanca, 1953 – Le Mans start of the race. *Jbbassibey.free.fr, Charles Plantivaux Collection*

[159] 12 Heures de Casablanca – The race begins and 0046M has yet to move out, the 6th car still parked on the left. *Jacques-Louis Bertin, Aston Martin: The whole story*

[160] 12 Heures de Casablanca – 0046M on track, in traffic behind number 3. *Cyril Teste*

[161] 12 Heures de Casablanca – 0046M (22) about to overtake a smaller class car. *Antoine Prunet*

[162] 12 Heures de Casablanca – Martignoni, probably after the race since he has no helmet. *Bruno Martignoni*

which finished 2nd overall, a most auspicious debut for Ferrari's answer to the 2-liter class. And with a 1-2 finish, Ferrari had established dominance on yet another continent.

The next month, January 1954, the Automobile Club Varese held its first meeting of the year, and based on their scoring system, Martignoni and Emilio Carnevalli tied for 1st as co-champions of the club for 1953.[154] However, Martignoni had been active through the year with other cars. He was also a consigliere of the Varese club.[155]

Two more races were scheduled in North Africa before the European racing season got underway and these were coordinated to be a week apart, making logistics simpler and less expensive for the entrants who were almost exclusively foreigners. The first of these was the Grand Prix d'Agadir for sports cars on February 27-28, 1954, and although Martignoni did not participate, perhaps he had intended to. He was a lifelong bachelor, living at 42 Via Cavour with his mother and stepfather, and family obligations might have intervened. His stepfather would die two months later on May 10 and could have had a health crisis that precluded Martignoni's travel to the first race.

Also missing at Agadir were Ascari and Villoresi who had left Ferrari for Lancia in spite of Enzo's effort to pacify them. Continuing his success was Farina who won again in a 375 Plus Ferrari (0386AM) on the 28th in the over 2-liter class, while the same 500 Mondial took the 2-liter class again on the 27th.

A week later, the Criterium Automobile du Sénégal, or the Two Hours of Dakar, was held on March 7. Martignoni did enter 0046M there as an independent, car number 44, and Scuderia Ferrari had Farina in the 375 Plus again, along with Maurice Trintignant in the 500 Mondial this time. Many of the participants from Casablanca were entered in the relatively small field of 17 cars. The course was 10.46 km, also mainly flat with an elevation of 8 meters and having a curious dumbbell shape. Known as Le circuit autostrade de Dakar, it used the north and south lanes of a divided four-lane highway as long, fast straights through what was then

a desert. Duncan Hamilton raced there the following year and described the course thusly:

> *It consisted of two straights – one with a slight kink in it – and two rounded 50 mph corners at each end. One way was slightly up-hill, the other slightly down. These two roads were separated by a narrow grass verge, so that if a car was to burst a tire and cross the verge the possibility of a head on collision with a car travelling on the other straight was very real.*[156]

Hamilton went on to say that top speeds went up to 190 mph (306 kph) and that although the race began at 5 PM, the temperature in the shade was still 90 degrees F (32 C).

[154] *Auto Italiana*, January 24, 1953, p. 46.
[155] Obituary, *Corriere della Sera*, July 11, 1986, p. 28.

[156] Duncan Hamilton, *Touch Wood*, p. 157.

This time, Piero Scotti won as an independent in his 375 MM and Farina did not finish, but the 500 Mondial proved itself again for Ferrari with a 2nd place finish. Martignoni finished 5th overall, but due to the classes being split at 1.5 liters, he was 5th in class, whereas he was actually 3rd among the 2-liter and under cars.

Fortunately, a color movie was made of the Dakar race and it has been posted on *YouTube*.[157] The full movie is 5 minutes, 47 seconds long and in two parts covers the 1953 and 1954 editions, the latter being in color and beginning at 00:01:49. The film skips through the race from start to finish, panning many of the cars at speed, none of which singled out are 0046M. All of the cars can be seen as the standing start is in progress, but the motion of the cars combined with the poor quality of the film and change in lighting make it difficult to identify the individual cars as the field accelerates past the camera. After tediously replaying the start over and over, frame by frame, I believe 0046M comes into view on the left side of the track at 00:02:41, having started near the rear of the grid. It is a hazy image, but it is recognizable as a small *berlinetta* and 0046M's signature feature of the big filler cap in the middle of the rear window is there, appearing as a bright spot above a larger light area that is likely the race number. The car I believe to be 0046M is not definitively red when it first appears and in fact looks grayish, but other cars look similar in the shadows at the start and display their true colors under better lighting. When the car in question moves into the distance, where the light changes, there is a faint reddish halo around the tiny image. At 00:02:53, there is an extremely quick look at what may be 0046M as it flashes by, and the color there is distinctly red.

The car looks dirty and worn after Dakar in **[163]**, having raced through desert dust and sand stirred by speeding cars. It looks forlorn, and with its purpose served for his Ferrari adventure, Martignoni abandoned racing it after Casablanca and Dakar, returning to his preferred Alfa Romeo 1900, which he raced exclusively from then on.

The Ferrari remained in Martignoni's ownership, however, and once it had returned to Italy from its sojourn in Africa

[157] *www.youtube.com/watch?v=6Jcg04CCw5M*, accessed February 24, 2018.

[163] 2 Hours of Dakar, 1954 – 0046M after the race with its African campaign over. *Martignoni l'africano by Robi Marzoli*

it was inspected in Varese on March 25, 1954, registered for road use on March 26 and given the rare booklet in Appendix 6 with the *licenza circolazione* of 39901 VA. Oddly, the booklet still lists the bodywork as a "spyder"– an open car. This was recorded at the Varese *Ufficio del Pubblico Registro Automobilistico* on April 7, 1954. It was in Martignoni's possession for a year and a half afterward and it evidently underwent a cosmetic makeover at some point, either to make it more suitable for his personal use or to make it more attractive for sale. Possibly, some of the work could have been done by Martignoni's own shop. Certainly, it was in the 1954-1955 period in Italy that 0046M was repainted in gray and Martignoni's choice of that color might have been influenced by current taste. Writing about the Zagato series of Fiat 8V cars of that day, Michele Marchianò says that, "The most popular color available was a dark metallic gray." [158]

Concurrent with the new paint, other changes were made. Polished aluminum trim was hand-crafted as an accent and fastened on with tiny screws. It extended from underneath the front turn signal lights on the body's edge, curving up to the top of the front wheel wells, then along the length of each side on the waistline lip that remained from the Touring *barchetta* body, over the rear wheel wells and terminating behind the rear fenders. Another strip of polished aluminum trim was attached to the ridge down the center of the hood. The side-mounted safety straps at the rear of the hood were removed and replaced with spring-loaded fasteners that opened with a quarter turn and popped up. At the rear, the thick seal around the fuel filler tube was replaced by two neat, semi-circular chromed plates that bolted around the tube through the Plexiglass rear window to hold a thin rubber gasket in place. Below the rear window, the Zagato name was removed, possibly to assert the craftsmanship of the shop that did the new work.

It might have been at this time that the firewall was covered with a thick layer of jute insulation. Zagato would probably have regarded its weight as a handicap and insulation was superfluous in *barchetta* form where the firewall was originally painted dark red. Otherwise, from testimony to be seen in the next chapter, the interior upholstery was left unchanged and the drooping headliner in the photograph from Casablanca was completely removed. This could indicate that the project was not finished, but there is no way to know.

Whatever Martignoni was doing with 0046M, he decided at last to part with it. From the last entry in Appendix 5, on September 13, 1955, he attested to the *Ufficio del Pubblico Registro Automobilistico* of Varese that the car had been *demolito* – demolished, scrapped – and it was removed from the register for that reason. The ruse was a common practice when such cars were to be exported out of Italy and were claimed to have been demolished, thus avoiding any further taxation.

Martignoni, though, continued racing and long-distance rallying, winning 1st in class at the 1958 XXV Mille Miglia and as late as 1970 completing a 9,000 km rally through the Sahara desert before retiring from motorsports at age 57.

As 0046M's days of serious competition come to a close, it is worth repeating what Massimo Giletti said about his late father and his friends because it captures the essence of that bygone era:

> *They were people who loved the rough races of that time, but also the dolce vita that surrounded them with beautiful women who were fascinated by these knights of risk, and the adventurous journeys in an era in which very few people traveled and they, the drivers, did it to reach tracks and circuits where they would challenge each other, but also where they would party, having fun together. How could you not succumb to such temptation if you run hard and you are twenty years old?*[159]

[158] *Le Zagato: Fiat 8VZ Alfa 1900 SSZ*, p. 56.

[159] *La Manovella Web*, No. 6, June 2021. Massimo Giletti is a popular television personality in Italy, as well as an heir to Giletti S.p.A.

Chapter 8

1955-1957: COMING TO AMERICA

0046M

I was working in my yard one weekend afternoon when my wife came out of the house with the cordless phone, saying that somebody was calling me about a Maserati. Since it was no secret that I owned a rare Ferrari, from time to time I had calls either offering to buy my car or sell me another, so I expected this to be another nuisance call. Instead, it was a stroke of luck. The caller was Don Minnick and he asked if I was the owner of 0046M. When I said yes, he went on to explain that he had bought an AG 1500 Maserati, number 083, and that among its paperwork was the shipping invoice from Italy – and my car was on it too! He said he had tried for six years to find 0046M and had only recently talked to Gerald Roush who told him about me. I wanted a copy, of course, which he said he would send, and I thanked him profusely for it and for finding me. It is included in Appendix 7.

The invoice was to Luigi Chinetti, whom I already knew had imported 0046M, but the details were enlightening. The date, September 9, 1955, was consistent with Martignoni's claim of *demolito* a few days later, and it reveals that Mario and Gastone Crepaldi, father and son Milan Ferrari dealers established in 1953, must have acquired the car from Martignoni.[160] They sent it to Chinetti from Genoa on a ship aptly named for a voyage to America, the SS Christoforo Colombo,[161] and the value declared for customs was

[160] M. G. Crepaldi Automobili S.a.s. at Via San Marco, 26, Milan. Now defunct.

[161] The SS Christoforo Colombo (Christopher Columbus) had only been in service for one year as a passenger liner sailing between Genoa and New York for the Italian steamship company Società di navigazione Italia. It was the twin sister ship of the SS Andrea Doria, sunk the following year in a collision with the SS Stockholm off the coast of Nantucket, Massachusetts.

[164] Luigi Chinetti in 1956 – importer of 0046M. *Revs Institute*

540,000 lire. In 1955 dollars, with tax, that was $870, which was the equivalent of about $10,240 in 2024. Of course the real cost was only known to Chinetti and Crepaldi, but certainly both made a profit.

Born in Italy near Milan, Luigi Chinetti came to the United States at age 39 in May 1940 as war was spreading in Europe. Having won twice at Le Mans, he already had a distinguished racing career and wanted to try Indianapolis. After a dismal result there, he turned to work as a mechanic and stayed until it was safe to return home. When he did so in 1946, the future was bleak there, but he met with Enzo Ferrari whose enthusiasm for his envisioned marque was infectious. On the day before Christmas, both men bonded and Chinetti was promised the exclusive Ferrari dealership for the United States, although there was nothing yet to sell. Chinetti continued his racing career and won Le Mans again, this time with a Ferrari, in 1949, which led directly to a number of sales for him.

The Chinetti Motors dealership Luigi formed sold and maintained new Ferrari cars, but he found a lucrative business in importing used Ferraris from Europe, particularly the outdated racing cars such as 0046M that could be had cheaply because they were no longer competitive. In the United States, meanwhile, the postwar interest in foreign sports cars was blossoming and there was a market for them. As the mystique of Ferrari grew, customers interested in exotic cars began to come around, but it seems finances were always a struggle and the business moved many times in New York, located for the moment at 252 East 61st Street when 0046M arrived in 1955.

One of Chinetti's regular customers was David Francis Leopold of Buffalo, New York. The son of a doctor father and artist mother, Leopold grew up in an affluent home and was prominent in social circles. His brother became a doctor and Leopold, too, was engaged in pre-medical education in the naval officer's college training program during the war. Post war he married, and for whatever reason he went into radio, possibly because he was endowed with a sonorous deep voice, or possibly because his brother-in-law was president of the radio station owned by the *Buffalo Courier-Express*, which his father-in-law published. He began as a news announcer on the air in 1947, eventually becoming vice president and then station manager of WEBR in Buffalo. Contemporary newspaper accounts show that he and his wife Rita were an active and sporting young couple, probably the "Scott and Zelda" of Buffalo, working on committees and presiding at functions both social and charitable. He was a golfer, hunter and fly fisherman; she was a trap shooter and skier of European slopes. Both were enthusiasts of elegant, fast cars. They were a glamorous pair.

When I found David Leopold's phone number in Stuart, Florida, I called him on July 29, 1988, and asked if he remembered John Sharpe, with whom I had talked earlier. "John *Reynolds* Sharpe," he corrected in the deep, precise and measured voice of a radio announcer, after which I told him I now owned the Ferrari that once belonged to Sharpe. Without a pause to consider, he said, "The serial number was 0046. In 1950 it was originally a spyder and Zagato put a top on it. I know because I owned it. I bought the car from Luigi Chinetti." Obviously, he had been told all about the car's origins by Chinetti, and I thought to myself, if I had only found this man first it would have saved me a lot of time.

According to Leopold, he was more than just a customer to Chinetti:

> *Luigi and I had a working partnership. Because I could get away with it* [importing cars] *without an excise* [tax]. *So it would be in my name and I just said to hell with it whatever you make with it is your money...Well it was totally legal. But he had to pay a ten percent excise* [tax] *and I didn't as an individual. Plus, he was one of the finest men I've ever known.*

Stan Nowak gave me Chinetti's home phone number later and I called the eighty-seven-year-old on September 12, 1988, to see if he had any recollection of 0046M, "the little *berlinetta* with the big gas cap in the rear window," but with so many Ferraris having passed through his hands, he had no memory of it. Although, he did remember David Leopold. In a soft, faint, accented voice he described him as, "a nice man. But he got in a big accident. I believe he was for a long time not himself, you know. He was hit in the head." When I told him that I had spoken to Leopold recently, he asked for his phone number and wrote it down. He said, "I tried to reach him years ago. He was a very fine man. All the nice people I like to remember."

The accident Chinetti referred to happened when Leopold was 31 years old, on September 19, 1955, in which he sustained facial lacerations and a broken jaw. When I asked him when he had bought 0046M, he told me, "If it was 1955 it was very late because I was smashed up in an accident and I had a 4.1 Mexico which I subsequently sold and then went down and talked to Chinetti and ended up with that car, so that could have very easily have been early '56. I'd lean to '55." The photograph of him from the *Buffalo Courier* **[165]** is dated November 15, 1955, and shows the scars on his chin and right temple.

Our conversations on all three occasions were recorded and are excerpted as follows; direct quotes from David Leopold are italicized:

[165] David Leopold in 1956 with facial scars from his 1955 automobile accident. *Archives & Special Collections Dept., E. H. Butler Library, SUNY Buffalo State*

That was a very inexpensive car at that point.

Me: How much did you pay for it?

Two thousand bucks, or twenty-five hundred.[162] *How much did you pay for it?*

Me: I paid fifteen hundred bucks for it.

Well, that's even better.

Truthfully, I have been away from any fine machinery since 1959. I owned five Ferraris,[163] *two SS-100's, an Aston Martin, a Maserati 4CLT Grand Prix car, a Talbot Grand Prix car, a Talbot coupe, Christ I don't know what else... a Bugatti. But you've got it pretty well knocked on your tracking down the history of that car. Is it still done in red naugahyde? I had it done. There was no headlining in it, there was nothing on the doors, the seats were covered in very cheap cloth.*

It was done in a rather dark metallic gray.

Me: Did you have it painted?

No, never had it touched. It didn't look at that time as if it had been redone, but of course I suppose if it had been redone in Italy, they're careful – it's not like things are done here.

Me: You say you redid the upholstery?

It was done by a man named Larry Jansen in Buffalo.[164]

The car ran beautifully. It was absolutely trouble-free. That little non-synchro gearbox is a sweet thing, if you know what the hell you are doing.

I asked him if he had any photographs from his time with the car, and he said there were many but he no longer had them, having lost all of his mementos in a bitter divorce from his second wife.

Leopold had a protégé in John Reynolds Sharpe **[166]**. After military service at the close of the war, Lieutenant

[162] In 2024 dollars: $23,540 or $29,430.

[163] 0226AT, 0046M, 0276MM, 0405GT and 0066M.

[164] Jansen Bros. Upholstery, 119 Broadway, Buffalo, NY. The seats were recovered and a headliner put in, both of the same red naugahyde, an artificial leather material.

[166] John Reynolds Sharpe when he was traffic reporter for WEBR. *Archives & Special Collections Dept., E. H. Butler Library, SUNY Buffalo State*

Sharpe entered civilian life at Buffalo University where he was president of his Sigma Alpha Nu fraternity. Fresh out of school in 1952 at the age of 26 and single, he went to work at WEBR radio as the night news editor under Leopold. No doubt the young Sharpe was impressed by his dashing boss and the procession of exotic automobiles that were a part of his lifestyle. Perhaps emulating him, or perhaps genuinely appreciative of foreign machinery, Sharpe became an enthusiast and his activities were intertwined with Leopold's, so he bought 0046M from him, around March 1956.

When I spoke to him on April 26, 1988, he remembered the Ferrari well and was quite talkative about his time with it, but as I saw all along, memory does not always serve us well after decades have passed. There are inconsistencies between what Leopold told me and what Sharpe told me, and both of them had omissions and errors compared to what I learned to be factual or logical. Leopold, however, I found to be the more accurate regarding the sequence of events, so I have taken what Sharpe said from that perspective.

Sharpe was the last owner to race 0046M, or more accurately to have it raced, at Watkins Glen on September 15, 1956, and two weeks later at Harewood Acres in Canada on September 29. Sharpe is not listed in the 1956 SCCA Membership Directory and thus had no competition license, but his and Leopold's mutual friend, Fred Woodmansee, is listed and had raced previously. He drove the Ferrari in both races and apparently there was some kind of partnership, since Sharpe commented that "Fred Woodmansee was in on it with me." I believe he meant as a collaborator, driver and mechanic rather than a financial partner. Woodmansee had raced against Leopold years before, both in MG TCs, and Leopold's nephew, James S. Leopold, remembers both Sharpe and Woodmansee well. They were family friends who were regularly a part of their social circle and his singular memory of Woodmansee is as a man with "big arms."

My two conversations with Sharpe are summarized by the direct quotes from him below, arranged to make a chronological progression:

> *Dave* [Leopold] *bought the car and he sold it to me, and I think he got wind of the car through Chinetti.*
>
> *I paid twenty-five hundred for it, I remember that.*
>
> *I had it maybe six, eight months ahead of that time* [i.e., the Watkins Glen race on September 15]. *I had it before I raced it a few months. Not too long though.*
>
> *That darned thing was a neat little car and I never had a problem with it. The thing ran in traffic and ran every place. It was a sensational little car. When I had it, it was gray with red leather interior. Dark gray, sort of a gun metal gray, darker gray than the normal gray. Charcoal gray.*
>
> *I used to drive the car every day in traffic.*
>
> *It had like an indoor/outdoor carpet.*

Races at Watkins Glen had originally been held on public roads, but in 1956 the organizers constructed a dedicated 2.35 mile (3.78 km) road course laid out on 550 acres. Paving of the new track, though, was actually only finished the night before practice, and when cars started to run on it, the surface began to disintegrate. Gordon MacKenzie, another of the participants, has been quoted as saying: "In practice the track was terrible. It was falling apart with every lap. Flying stones were everywhere. Harry Carter and I had our windshields busted by the rocks. Carter's arm was injured by flying debris because of his habit of hanging his elbow over the side of the door."[165] On top of that, the Sports Car Club of America withdrew their sponsorship and sent a telegram urging the drivers not to compete. The consensus of the drivers, however, was to race and not to cheat the crowd of 25,000 who had come to see them, so the events were shortened and the races went on.

Bill Green, the historian for Watkins Glen mentioned in Chapter 3, told me that an audio recording of the race was made and a record album produced from it, which I found. The album was by Folkways Records and Service Corp. in 1956: *Sounds of the Annual International Sports Car Grand Prix of Watkins Glen N.Y.*, FPX 140. Side 2 has highlights from the Grand Prix race, which was the event Sharpe entered, and the narrator points out the high-pitched scream of 0046M's 2-liter V12 when it passes, but he misidentifies it as a 166 Inter. Both Leopold and Sharpe called my attention to the record as well. Another record album was produced by Riverside Records from the 1956 Watkins Glen and titled, *Sports Cars in Hi-Fi, The World's Greatest Competition Cars:*

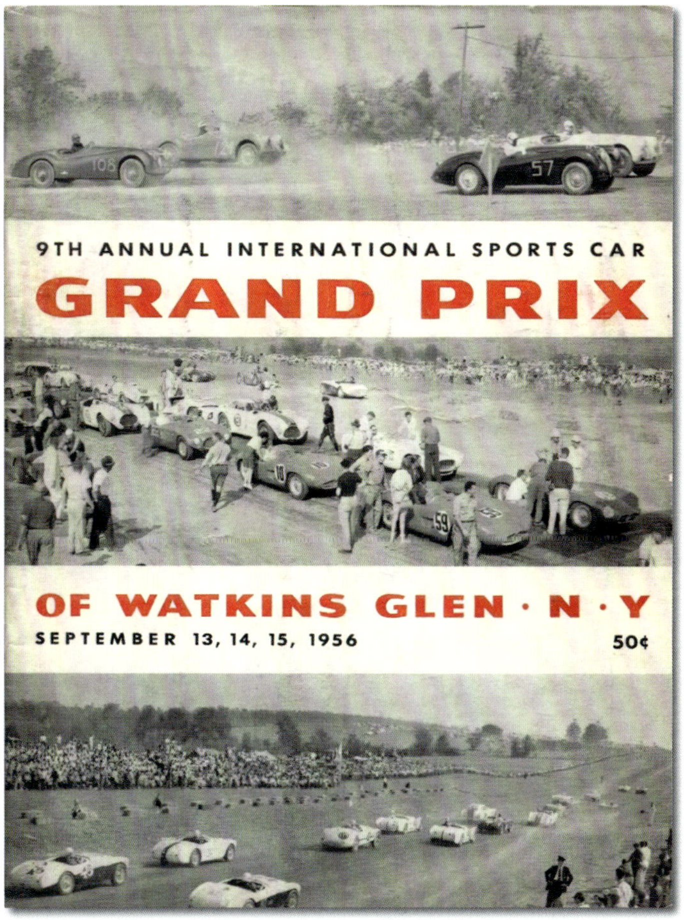

[165] *www.vintageracecar.com/watkins-glen-1956/2/*, accessed March 10, 2018.

[167] Watkins Glen, 1956 – Pre-race photograph of 0046M in its dark gray color. *Fred Leydorf*

at Rest and at Speed, RLP 5002. Side one records the sound of the cars individually at rest and track 16 is the sound of 0046M, again called an Inter. The jacket dramatically exclaims, "...what makes the 166 Inter Ferrari sound like three cars – all of them ready to fall apart? Twelve tiny cylinders (six against six), twenty-four valves, mashed open by that steep cam, snapped shut by those double springs, fifty times every second."

Green was also able to provide me with a copy of the Biographical Data Sheet for Woodmansee. At the time, he was 51 years old, lived in Buffalo and was a machine shop owner. He had raced three times previously. Under "Noteworthy facts about you or your car" there is the comment, "Recently imported – has not competed in this country to this date." The sheet also gives the car's race number, 87. Green explained to me that the grid positions were drawn by lot that day, probably by class, and a diagram of the standing start shows 0046M on the 6th row, grid position 15, for the feature Grand Prix race of 19 cars.

As I mentioned in Chapter 3, Fred Leydorf was there that day and captured a photograph of 0046M in full color **[167]**, and clearly it is a dark gray with the new accent trim down the hood and along the sides. There is protective tape prominently applied to guard against flying pavement debris.

Sharpe breathlessly related his account of Watkins Glen as follows:

> *An' when I got the car, the interesting part was at Watkins Glen that first year the Ferrari mechanic came up from Chinetti, an' they sent the whole crew up. An' they came over on that little car an' they hadn't seen it. They didn't even know I had it an' they tuned it for nothing 'cause they didn't want any Ferrari to run badly at Watkins Glen.*
>
> *We taped over the front of the car 'cause of the simple fact the track was soft an' it wasn't until about an hour before the race that SCCA even gave permission to have anybody run on it an' everybody taped their cars because the track was flyin' up all over the place. That's why it was taped. In fact the Ferrari team taped it for me. They had the first Testa Rossa that was in this country.*[166] *They brought it to the track that day an' they came over to see my car an' the mechanics from Chinetti came up an' they tuned up our car for us for nothin' an' everybody taped 'em because you know it was a nice car an' it was knockin' the hell out of it an' as a matter of fact in that race a chunk of blacktop thrown up by uh Duncan Black was drivin' a horse the 4.9* [4.5 (0382AM)] *Ferrari pulverized our windshield an' we had to drop outa the race – it just went milky, you know, an' we had to get the hell outa the race. The whole track came apart that day. That's why it was taped.*
>
> *That same race after that happened he jammed an' couldn't see anything an' that's when that rear half spindle spinned, on the right rear I think it was.*
>
> *We spun a wheel hub when we were running second in class an' ahead of the Testa Rossa, an' at that time, I got married an' I was a little short of money so I sold it an' I can't remember who the hell I sold it to. I really don't remember. An' got an MG TC an' some cash in trade an' that's the last I heard of the car.*
>
> *An' of course we not only spun the wheel, we got the windshield buggered up, an' ah I'm sure it was the right*

[166] The Testa Rossa that raced that day was 0614MDTR, owned, driven and crashed by William Helburn. As Keith Blumel confirms in his *Cavallino* No. 230 article on the car, it was indeed the first Testa Rossa in the US, and it was sold to Helburn by Chinetti Motors. Since no great distance was involved, it is possible that Chinetti mechanics accompanied the car. A photograph (not shown) of Helburn's wrecked Testa Rossa after the race has Luigi Chinetti either removing or replacing the hood, so he was there in person. Earlier in the conversation, Sharpe had alluded to its being a Testa Rossa that had run at Le Mans, but he had conflated 0614MDTR with another Testa Rossa that had been entered but did not show up, 0642MDTR, which had recently been sold by Chinetti to Robert Publicker. Chinetti had also entered two cars for the race, which did not show up. As for Chinetti mechanics taping up the front of 0046M, the photograph of Helburn's wrecked Testa Rossa shows that its front was not covered in tape as 0046M's was. Photographs show that several other non-Ferrari cars were fully taped, so it is a moot point.

rear wheel but we did spun a hub. We had to quit the race because we spun a hub.

We had a trailer we towed it back on.

I'm tryin' to think – it may have been that we were going to have trouble getting the hub for that car an' somebody wanted it or we said why don't we try to sell it.

Whoever bought it, bought it before we fixed the spun out wheel hub. I know that.

I sold it locally. I'll be honest about it, I don't remember who I sold it to or even where I sold it, I'm not gonna be much help.

What Sharpe refers to as a "spun wheel hub" is a broken rear axle, the same failure that caused Bosisio to withdraw from the Mille Miglia and a common problem with early Ferraris. He notes that it was the right rear axle shaft that broke, a fact that will become significant later. But, Woodmansee's natural reaction to the "pulverized windshield" and suddenly driving blind would have been to slow down and pull off, so it is highly unlikely that 0046M suffered two catastrophic, race-ending events in the same race. Leopold's account of what happened at Watkins Glen was different from Sharpe's and more accurate. Returning to that conversation:

Me: He [Sharpe] said the car had broken a wheel hub during the race, during the Watkins Glen.

No somebody, somebody in front threw something, I suspect a stone, and the windshield became absolutely opaque. Totally – and that was it.

Strangely, neither Leopold nor Sharpe ever mentioned the second race – at Harewood Acres in Canada on September 29 – and since I did not know of it at the time, I did not ask about it.[167] It was the second race held in 1956 at a new 1.9 mile (3 km) airfield track location for the British Empire

INDIAN SUMMER TROPHY RACES

PROGRAMME OF EVENTS

FRIDAY 28th OF SEPTEMBER

9:00 A.M. - Registration Starts
9:30 A.M. - Safety Inspection Starts
12:00 Noon - Entries Close
12:00 Noon - Official Practice Starts
3:00 P.M. - Registration Closes
4:00 P.M. - Safety Inspection Finishes
5:00 P.M. - Official Practice Finishes
5:10 P.M. - Compulsory Drivers' Meeting

SATURDAY 29th OF SEPTEMBER

8:00 A.M. - Official Motorcycle Practice Starts
10:00 A.M. - Official Motorcycle Practice Finishes
11:00 A.M. - Start - Harewood Trophy Race Sport Cars Under 1500 C.C. - 33 Laps
1:00 P.M. - Start - President's Trophy Race Motorcycles - Junior - Senior - Expert - 33 Laps
3:00 P.M. - Start - Indian Summer Trophy Race Sport Cars Unlimited - 33 Laps
8:00 P.M. - Victory Banquet and Presentation of Trophies.

[167] A third race is listed in *Northeast American Sports Car Races 1950-1959* on October 7, 1956, at Thompson Raceway, Thompson, CT, Race 6. However, this is an error. The results listed for Race 6 are the results from the Watkins Glen Grand Prix race.

Motor Club, just above the north shore of Lake Erie, only 60 miles (97 km) from Buffalo. The club historian confirmed that 0046M was there, number 65E, entered by Sharpe and driven by Woodmansee. Furthermore, Leopold was there also, entering and driving his next Ferrari, 0276MM, a 250 MM and another racing veteran. In a remarkable coincidence, it was a reunion because the two Ferraris had been on the same team before, four years and 6,000 km distant at the 12 Heures de Casablanca. 0276MM had been Luigi Piotti's car, but it was unrecognizable now in a new body. Unfortunately, the BEMC has no record of the race results, but the program shows that there were two sports car races that day, interspersed with motorcycle races, and both Woodmansee and Leopold were listed for the 33-lap Indian Summer Trophy race at 3 PM for cars over 1.5 liters. There were 37 cars entered in that race, which began with a Le Mans start.

Frank Smart, an eighteen-year-old newly arrived émigré to Canada, was there that day and photographed 0046M in the paddock. His photographs **[168-172]** show the car with the flat glass of the right windshield replaced and parked side by side with Leopold's 0276MM. The front end of Sharpe's car is still wrapped in tape as it was at Watkins Glen and Leopold's car has been wrapped in an identical manner, illustrating the alliance between the two. The photograph of the engine compartment reveals where the windshield washer reservoir was located, since its remaining bracket can be seen wound with cushioning on the right side, near the firewall. The arcing tube nearby was probably the supply line. The front and rear photographs of the car on grass could have been taken after its retirement, and James and Jonathan Leopold, both nephews of David Leopold, recognize the man standing at the open passenger door as John Sharpe. On the rear, the car bears the New York State license plate 83-48BL.

Without doubt, the sequence of events Sharpe described of a broken axle leading to a retirement in its final race and the subsequent sale of the car must have begun at Harewood

[168] Harewood Acres, 1956 – The engine of 0046M. *Frank Smart*

[169-170] Harewood Acres, 1956 – 0046M and Leopold's 0276MM sit together in the paddock, taped alike. *Frank Smart*

[171-172] Harewood Acres, 1956 – The Ferrari off the track on grass, likely after the rear axle broke. John Sharpe stands by the open passenger door. *Frank Smart*

Acres and not Watkins Glen.[168] And the person whom Sharpe was unable to recollect as the buyer was none other than Leopold again. While he did not explain the true nature of the situation – which was probably that he got the damaged car back as a steal or merely brokered a deal for it – Leopold was clear on the progression of ownership, as seen in our conversation:

> **Me: So, that's an odd thing. You bought the car from Chinetti, and then sold it to Sharpe, and then bought it back from Sharpe again.**
>
> *That's because it really wasn't for him. I took it back from Jack because it was not really a practical thing for him. I, in turn, sold it to Delavan Lee in Detroit. He was a dealer in used sports cars. Very shrewd.*
>
> **Me: Do you recall the name of the business that Mr. Lee had in Detroit?**
>
> *It was something like World Sports Cars. Delavan Lee was a shyster.*

For Leopold to have found a ready buyer in Delavan Lee would not have been unusual. The sale of his 340 Mexico coupe that he alluded to earlier as having occurred in 1955 was actually before his accident and it ended up with Lee in 1954. Turning over cars as often as he did, Leopold was likely considered a good customer by Lee, as he had been by Chinetti.

Continuing its travels, 0046M moved on to Detroit, probably trailered there, given its broken axle shaft.

Delavan Smith Lee had a reputation as a flamboyant Detroit car salesman with a taste for fast cars. Pre-war, he had been a salesman for an advertising agency, but after his active duty service in the Navy as a Chief Warrant Officer **[173]**, he turned his sales skills to the post-war automobile boom. When he was a salesman for Louis Deeb Cadillac in the early 1950s, he raced his own J2X Cadillac-Allard, which he pulled on a trailer with his Cadillac convertible, all

[173] Delavan Lee, Navy service photograph. *Lee Jones*

of which were painted metallic blue. In road races and also ice-racing on Lake Orion, Michigan, he was a contender but rarely a finisher. His only two wins were at hillclimbs, but he did manage 20th place at the first 6 Hours of Sebring race in 1950. In 1954 he sold the Allard,[169] and at age 40, he was Sales Manager at Lawrence Falvey's import dealership. One of his salesmen recalled, "Del tried to be smooth, but he generally came on like Attila the Hun in flannel slacks and a cheviot [wool] jacket. His speech was both garrulous and foul and he was no respecter of persons."[170] His nephew confided to me that his mother, Del's sister, was of the same opinion. Lee moved on to a Jaguar XK120M and it was not long before he

[168] Leopold also dropped out of the race with a mechanical issue. He said he was happy to do so because the pace was frightening.

[169] Lee bought Allard J2X number 99J 2123 in August 1951 through Wood Motors of Detroit. It was sold to Fred Lavel who put in a bigger DeSoto engine, re-bodied the car, and raced it at Bonneville Salt Flats. *https://rmsothebys.com/en/auctions/mo20/shift-monterey/lots/r0086-1951-allard-j2/951500*

[170] David E. Davis, Jr., "Remembering Del Lee: Allards, Ariels and Atilla the Hun," *Car and Driver*, May 1978. He uses the incorrect spelling "Delavan."

struck out on his own to open the grandiosely named World Sports Car Center.

To learn more about Delavan Lee, I turned to Dick Merritt, who had come to Detroit straight out of college in 1956 to work for Ford as an engineer and later was one of the founders of the Ferrari Club of America. I had already consulted him in my search for a transmission, and when I called him again on October 9, 1988, Merritt brought up the likelihood of another man, Richard "Dick" Londergan, being involved with 0046M in Detroit. I quote Merritt at length because of his direct knowledge and assessment of the situation:

> *I went to Detroit in June of '56. I got out of college in Colorado and went to Detroit and was of course pumped up from reading* Road & Track *and so forth and I got acquainted with everybody in the Detroit area who was interested in exotics. And I used to go down to Del Lee's place literally every Saturday just to see what great things were there. And soon I met Londergan who was just around the corner.*[171] *Londergan was on Fenkell and Lee was on Livernois.*[172] *And Londergan was probably the only mechanic as such that was really intrigued and interested in futzing with those things.*
>
> *Lee was not the kind of guy that really wanted to do a lot of work on a car. He wanted to get it in there, put it on the showroom, sell it, get rid of it. And he had guys who would certainly wash 'em and polish 'em, but as far as doing any real work on 'em, I don't think he really had much in the way of anybody around that was a mechanic as such. It would be much better for him to farm the car off to Londergan for a tune up. Lee certainly didn't want a high-paid mechanic on the payroll. He really just had a showroom.*
>
> *Let me tell you, he was a pretty big operator. He was wired in to Chinetti, and if Chinetti couldn't sell an old Ferrari, it would end up in Detroit with Del Lee. And there was all sorts of stuff went through there.*
>
> *Londergan was a fairly decent mechanic and I suspect Del Lee was using him as a mechanic-dash-salesman. As in, if you can con a guy into buying this thing please do. Lee was a city slicker – he was a button down shirt, nice tie, blue blazer type. He was a pretty spiffy guy; he was a dapper dude, absolutely a dapper dude. Londergan was absolutely the opposite. He was just about the greasiest slob you've ever met, but eccentric and somewhat likeable. And I think Lee was smart enough to realize that if people didn't gravitate to Lee's personality – which a lot of people did – they might gravitate to Londergan's personality, which was really the opposite extreme.*

Evidently then, Lee bought 0046M then turned it over to Londergan to repair, certainly not wanting an inoperable car in his World Sports Car Center showroom. That the axle was still broken I infer because Fred Leydorf told me he saw the Ferrari sitting outside a shop on Findley St. in Detroit in early 1957 and recognized it from Watkins Glen. He stopped in to ask about it and was told that the car was waiting for parts to come in.[173] Findley is a residential street and Leydorf surely confused it with Fenkell where Londergan's Dick's Auto Repair was located. Merritt, though, said he never saw the car, which might be explained by its wait at Londergan's shop rather than being in Lee's showroom, which he frequented. Furthermore, it might have come and gone from Detroit before he became an insider.

Both Lee and Londergan are long deceased, but in 1992 I was able to track down Lee's son, Michael, in New Orleans where he was involved in the sales of Donzi speedboats, having moved there when his father relocated his car

[171] Actually, Londergan's business was about two miles (3.23 km) away at 7144 Fenkell Ave.

[172] 17333 Livernois Ave.

[173] Leydorf said his interest in Ferraris grew and he kept hoping to find 0046M again, but when the opportunity presented itself to buy something similar that reminded him of it, he did, purchasing 0066M, a 166 Mille Miglia *berlinetta*, in 1964 – his first Ferrari. The car had reached him after passing through several hands and having been imported by Chinetti in 1958, from whom it was first purchased by his good customer, David Leopold. Later on, Leydorf sold 0066M to John Bond, who in turn eventually sold it to Tohru Horinouchi along with 0020M and a spare engine and the transmission that would finally become a part of 0046M. The world of old Ferraris is complicated but small.

business. His memory of World Sports Car Center in Detroit was hazy, but he had a vague recollection of a car with a big gas cap in the rear window.

Londergan eventually received and installed the new right side axle shaft and it was time to put 0046M on the market. In some publication which I have yet to find, 0046M was advertised for sale and Charles "Chuck" Saffell of Huntsville, Alabama, saw the advertisement. A deal for the purchase of a 300SL Gullwing Mercedes had just fallen through, so he and his wife, Audrey, packed their bags and drove their white TR2 Triumph to Detroit where they traded the TR2 and some cash for the Ferrari. Decades later, Audrey recollected that the total price they paid was about $3,000.[174] Chuck recalled that there were two other Ferraris there. One was a 340 Mexico *berlinetta*, which he recognized because he had accompanied Ernie McAfee to the 1952 Carrera Panamericana, or so he said.[175] That car, he said, was no longer red as he had seen then but was a dark color. 0226AT was the red 340 Mexico that David Leopold had owned in 1954 and it had been repainted black. That car was in the Detroit area in 1957 and passed through Del Lee's hands a second time the next year, in 1958 when he sold it to Londergan, so it is quite possible that 0226AT was at Lee's showroom that early.

Hopefully, whatever bags Chuck and Audrey brought along were small ones because the only places to stow anything in the Ferrari were either in the space above the spare tire behind the seats, which was scant, or in the passenger's lap or foot well, which was uncomfortable. In cramped quarters, they set out on the 645-mile (1,038 km) drive back to Huntsville at a time before interstate highways.

---

[174] In 2024 dollars: $33,680.

[175] Three 340 Mexico Ferrari *berlinettas* were built for the 1952 Carrera Panamericana: 0222AT, 0224AT, 0226AT.

Chapter 9

1957-1966: SOLD DOWN SOUTH

0046M

Chuck Saffell was a product of Southern California and was infused with the car culture there in the '50s. He was a friend of Ernie McAfee and in his self-aggrandizing way, Chuck liked to say that he was his co-driver in the 1952 Mexican Road Race, but history does not bear that out. Instead, he could have been with him as part of the team. Mechanically inclined, he was an engineer and worked in the early missile programs. In 1952 at age 34, he married Audrey Devine, who was 17 years old at the time, and in 1953, they moved to White Sands Proving Grounds in New Mexico where a photograph shows them with their XK140 Jaguar, their first in a long string of exotic cars. They moved to Huntsville, Alabama, in 1954 where Chuck applied his talents at Redstone Arsenal on missile design and became active in the Alabama Chapter of the Sports Car Club of America in Birmingham.

After I located Chuck in 1988, I had six telephone conversations with him over nearly a year's time and one of his letters is reproduced in Appendix 17. He was 70 years old then and he and Audrey had divorced long before in 1962. Chuck was very talkative, and where time had not dimmed his memory, he liked to embellish his recollections, sometimes leaving me to separate fact from fiction. Audrey confided to me later that her father had said Chuck "told the truth attractively." But he never mentioned Audrey, and hearing him tell it, I assumed that the story was all his. That was a mistake.

Ten years after I spoke with Chuck, I went to the *Cavallino* Classic Ferrari Concours in Palm Beach, Florida. At a luncheon, John Barnes, Jr., then publisher of *Cavallino* magazine, invited me to sit at his table since he knew me as a contributor. When Barnes introduced me to the others at the table and mentioned that I owned the Zagato 166 MM, the man across from me, Michael Lynch, told me he had interviewed Audrey Saffell while researching his book *American Sports Car Racing in the 1950s* and that she had mentioned owning my car. It turned out that she had played a more significant role than I thought, so Lynch put me in touch with her.

Audrey was as colorful a character as any who ever owned 0046M. She was a sports car enthusiast just like Chuck, and it was together that they had embraced the hobby. She entered her first race at El Paso, Texas, while they lived in New Mexico, and they ran an MG TD at a race in New Orleans in July 1955 after moving to Alabama, with Chuck taking a 1st in class. Then, according to Chuck, he organized the first SCCA race at Courtland, Alabama, on an unused World War II airfield, which was held September 4, 1955. There in the MG, Chuck took another 1st in class and finished 3rd overall. They added a Triumph TR2 to their stable in 1956, but Chuck crashed it at the SCCA Nationals at Walterboro, South Carolina, in March 1956. Doing her part for the team, Audrey came through in the MG, racing it twice and placing 6th and 22nd, one week after her 21st birthday.

In June 1956, the Saffells bought a 1953 Ferrari 250 MM Vignale spyder, 0348M, complete with trailer, spare parts and spare tires for $4,000[176] and went racing in earnest under the name Scuderia Furioso. In July, Chuck was 2nd overall and 1st in class at the second Courtland race and

[176] In 2024 dollars: $46,400.

[174] Chuck and Audrey Saffell display trophies in the 1957 Ferrari Yearbook. *Audrey Saffell Sundin*

[175] Audrey Saffell Sundin and her racing trophies. *Audrey Saffell Sundin*

Audrey won the ladies' "Powder Puff" event, as women's races were often called. They continued racing the 250 MM through 1956 and 1957 very successfully, especially Audrey, who usually won the women's races. A photograph that appeared in the *1957 Ferrari Yearbook* **[174]** shows them both with 1st place trophies from that year, and in the photograph above, Audrey poses with more of her trophy collection in 2010 **[175]**.

I estimate that it was early in 1957 that the Saffells decided to become a two-Ferrari family and made the trip to Detroit to buy 0046M. A3 Chuck told me:

> *It's a funny thing, I saw this ad in Detroit for a Ferrari for sale, so I gave 'em a call, they said they had it, I went up there and looked at it, decided I wanted it, ahh... I had just paid off an airplane – a big one – from Bank of Detroit, and I financed it with Bank of Detroit and left with it.*

The airplane is dubious and, typically, Chuck omitted the trade-in and the fact that Audrey was with him, but she was able to elaborate with more veracity on the transaction:

> *We had a Triumph which we took there and traded it. It was at a dealer where we bought it. What happened was we wanted to go across to Canada and we couldn't because they had the registration to the Triumph that we were trading. So that night we couldn't go across. They already had all the paperwork on the Triumph and we were gonna do the swap in the morning, you know. But we did buy that from the dealer in Detroit.*

> *I can visualize the dealership. In my mind right now I can tell you that the dealership faced east. Isn't that strange?*

Del Lee's building at 17333 Livernois Ave. faced east.

When they left Detroit, the car had the 1957 Michigan license plate EG 5037 from Wayne County where Detroit is located. Records no longer exist about that plate, but at that time, Michigan had a plate-to-owner system, not plate-to-vehicle. Once a vehicle was sold, the plate no longer would be valid for use on that car and would only be legal on subsequent vehicles titled by the original licensee after being transferred. It is unlikely that there was an intervening owner of the Ferrari in Detroit, since that hypothetical owner would have removed the plate as it was his property, usable on another car, and the plate could not be transferred to Chuck and Audrey. The most likely explanation for the Michigan plate is that Chuck and Audrey registered the Ferrari to themselves in Detroit after the sale in order to have a legal plate while traveling back, and they could have used the local address of World Sports Car Center. That they kept the plate on the car the remainder of the year in Alabama says that it must have been registered to them.

Both Chuck and Audrey had vivid memories of the trip back from Detroit and Chuck rhapsodized on the open road:

> *I drove that God damned thing all the way from Detroit to Huntsville, Alabama, over open roads with four open pipes on it, scaring horses all the way through the whole state of Kentucky and every place else. I got stopped in Louisville, Kentucky, for making noise and told them I was just ferrying a car. Very memorable experience driving down through that beautiful horse country about 120 miles an hour* [193 kph] *with four straight pipes.*

Audrey, on the other hand, recalled more of the discomfort:

> *When we bought that car the only interior in it was purple flocking – well it might not have been purple but it wasn't black and it wasn't gray. And I had that dust, that flocking in every garment I owned. I mean my bra was lined with flocking and underneath the seats it was an inch thick. They'd just gone in and sprayed everything, you know, with this flocking because all it had was the two red seats. And they must have done it for resale, you know, right away. Boy, it was a mess by the time we got home from Detroit.*

I am not sure what the "flocking" Audrey referred to was. The car had fiber mats that were a faded black color and Del Lee's minions might have sprayed them with paint or something else to give them a fresh look. Also, there was jute fiber insulation on the firewall. With the flip-out side vents open and the window vent wings and rear windows propped open, there would have been a strong air flow coming up from the foot wells and with the speed they were going, that probably blew detritus up and all over them. In any case, there was no sign of any purple color in the car. The red headliner David Leopold had installed was still in the car, and when I removed it years later there was no sign of the bare aluminum underside of the top ever having been painted.

The trip was not without consequence, either, since Audrey told me that they began hearing a noise from the rear late in the journey. Chuck was irate:

> *I don't know who that son of a bitch was I bought it from. I could have killed him if I'd got my hands on him. Because when he sold me the car it had a broken rear end in it.*

As previously noted, a rear axle had broken with Bosisio in the Mille Miglia, although he did not specify which side, then the right side had broken with Sharpe at Harewood Acres. The axle shaft is encased in a thin-walled tube with flanges on each end that mate on the inside with the differential housing and on the outside with the brake backing plate, illustrated by the photograph of the left side axle tube and differential housing of 0046M **[176]**. At each end of the tube are bearings supporting the shaft. When the axle shaft breaks, the loose ends spin uncontrolled inside and can damage the tube, which is likely what eventually happened, possibly twice. Del Lee knew the shaft was broken, so that was the part the car was waiting on at Londergan's shop,

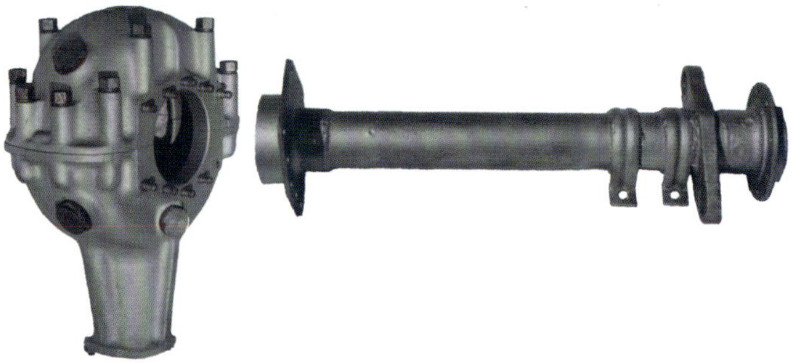

[176] The differential and left rear axle housing of 0046M. *Author*

and Londergan probably waited to pull off the axle tube and retrieve both ends of the broken shaft until after the new axle came in, since doing so would have left the car immobile. When he did, he either could not see the internal damage in the tube that would lead the tube to crack under further use, or if there was already a crack, he must have decided to let it go as-is rather than endure another long wait. As Chuck drove along, a crack either developed or persisted and oil leaked out, the bearings began to run dry, and that was the noise they heard.

Chuck said – and Audrey confirmed – that he had befriended Romolo Tavoni, the Ferrari team manager, in 1957 at the Sebring race by acting as a go-between with the race officials. That connection served him well, or so his story went:

> *Tavoni did one thing for me which was beautiful. The right axle housing on the rear – ok, up against the differential is a flange then there is an axle housing that goes out to the brake assembly. It developed a crack. And God bless him, he got me one for free, the last one that was left.*

What he got must have been a set of both the right and left axle tubes, which are not interchangeable due to the angle of their leaf spring mounts. He used the right side one for his repair and kept the left side, passing it on along with the car. That finally explained why there was an extra left axle tube when I bought the Ferrari.

Talking to Chuck, you often got several versions of the same story, and in one Chinetti was the source of the part, which might have been the case since Audrey told me, "We had to get everything out of Chinetti," but Tavoni was more frequent. One tale that never varied over many repetitions was his spurious version of the "history" of 0046M and I feel certain that it came from Del Lee's imagination as part of his sales pitch. According to Chuck:

> *The car had originally been built by Ferrari for Pietro Taruffi to run in the Mille Miglia. It was a one-off car; they only built one like her. The car had some kind of a problem and I don't think finished that race. Then a Frenchman bought it and painted the damned thing blue – French racing blue. The car was entered at Le Mans and sad to say it hit a dog on the first lap and retired.*[177]

Neither Chuck nor Audrey could pinpoint the date they bought 0046M, but I can make an estimate from two of their photographs. The first **[177]** looks like the sort people take when they proudly record the purchase of a new car. The trees are completely bare of leaves, indicating winter has been well underway, and the area in front of the fence is clear of vegetation. Even so, Chuck and Audrey are not dressed as though it is very cold, so winter is waning. In a companion photograph **[178]**, their son Arthur is posed sitting on the car's fender and is dressed more warmly in a sweater. Arthur was born October 28, 1954, and he looks to have the size and appearance of a child two years and a few months old. Taken all together, these clues lead me to believe they bought the Ferrari around late February or early March 1957. Since they had planned a trip to Sebring for the March 23rd race, they might well have thought to use the Ferrari as transportation there, but that was not to be with the cracked axle tube. In that sequence, when they did get to Sebring, Chuck could have mentioned his need for the tube once he had befriended Tavoni and used that connection to his advantage.

[177] Part of the story seems to have been borrowed from another 166 MM imported by Chinetti in the mid-1950s, 0068M, which was painted blue by the Frenchman Lucien Farnaud and raced at Le Mans in 1951, but it retired from engine failure on lap 65, no dog involved. This was the 166 MM later purchased by Gerald Roush mentioned in Chapter 3, and it was at the Barchetta Reunion with its then owner, David Cottingham, painted blue.

Audrey, it turned out, was the principal driver of 0046M in the Saffells' tenure, and she loved it. When I asked if either of them had raced it, she replied:

> *Technically, no, we did not race that car. I sure did put it to the coals living on the mountain in Huntsville, that five miles* [8 km] *of mountain road. Oh, it handled magnificently!*
>
> *I think of it so often because that was such a unique little car. That crashbox was a bear. If you missed a gear, you better just stop and start all over again.*

In her daily use the car was hit once in Huntsville. I noticed a dent in the left rear fender in one photograph **[179]** and I asked Audrey about it. She recalled:

> *Now that you mention it, it did have a dent, in the fender.*

[ABOVE: 177] Chuck and Audrey pose by 0046M, probably soon after purchase. *Audrey Saffell Sundin*

[BELOW: 178] Their son Arthur, age 2, posed on the fender of 0046M. *Audrey Saffell Sundin*

[179-180] 0046M by the Saffells' garage, engine and seats removed. Note the dent in left rear and the green grass and weeds indicating summer. The hole for the third wiper post has been filled in and the dark gray paint will be changed to red. *John Apen (Charles Saffell)*

It was a lady, a little old lady, elderly lady in Huntsville. She was backing out and I honked at her and she just kept backing. Yes, I thought it was the left fender. It wasn't any major accident. I was caught behind her – she was facing one way and I was facing the other way, and she just crunched into me.

While the grass was green, leaves were on the trees, and weeds were grown up along the fence by the garage, Chuck began a restoration on the Ferrari. There was the dent in the left rear fender that needed to be repaired, so why not repaint the whole car a proper red to match their 250 MM? While at it, why not rebuild the engine? And so it went.

The photographs in **[179-181]** and **[183-186]** show the sequence of the work. On the front bumper of the step van they used for a tow vehicle can be seen an advertisement for the July 7 race at the Courtland airfield, so Chuck began either around or past that time. In the background can be seen the changing of the seasons as the trees lost their leaves and the grass turned brown, so the restoration was a slow process. Note their red Vespa with the Ferrari badge, good for getting around the paddock on race day. These photographs are from slides in the collection Chuck sold to John Apen and the color on the film had shifted over the years toward red. Here, the film's color has been restored as well as possible, and before painting the car's color appears consistent with the dark metallic gray both Leopold and Sharpe described and as seen in the Watkins Glen photograph.

Chuck told me that the engine had a lot of "blowby" (oil smoke) and that was the reason he rebuilt it. He said that he had ordered a head gasket set from Chinetti Motors:

When I rebuilt that engine, Luigi Chinetti was still the

[181] The combination head gasket arrangement on a 166 engine. *Author*

[LEFT: 182] Engine 0046M on Chuck's homemade stand. *Audrey Saffell Sundin*

[BELOW: 183] Chuck has the rebuilt engine on a hoist, ready to lift it off the stand and roll the now-red Ferrari under it from the garage. *John Apen (Charles Saffell)*

[184] Car and engine together again. *John Apen (Charles Saffell)*

Ferrari distributor in New York. And I knew Luigi real well as we raced against each other in Mexico and he sent me a set of head gaskets. And you know that car uses little funny brass rings around the pistons, around the cylinders, and a funny rubber gasket around the outside **[181]**.[178] *And the first time I reassembled that engine, he sent me the wrong thickness of gaskets. And I coulda killed that wop. Oh God, you know, I had to tear it clear back down again and do it from scratch.*

In photographs **[179]** and **[180]**, Chuck has removed the engine, doors, steering wheel and seats to restore the car. The grimy engine is supported in the garage on a stand he constructed himself for disassembly **[182]**. Evidently, he fully dismantled the engine because he referred to having taken photographs of the internal parts laid out on a table. Sadly, those photographs are lost. In **[183]**, Chuck is lifting the now shiny and rebuilt engine off the stand with a hoist, preparing to roll the now red Ferrari out of the garage shadows and under the engine to reunite them, ending with the engine back in the car **[184]**. In **[185]**, both Ferraris are in the picture with 0046M being reassembled and **[186]** shows the finished project. In spite of delays, the car was together again, but Audrey had her reservations about the job:

I know I had to put a heater under it at night in the wintertime and a blanket over it to even hope to get it started in the morning because there wasn't enough compression. Once you got it going, fine, no problem. When he did the work and at that time he was very, very capable, he didn't do it completely because he didn't have the money. He didn't rebuild it as much as he should have. That was a big problem. That's why we had so much [trouble] *starting it in the cold weather,*

[178] "A separate L-profile steel ring was located in the gap between the block and liner. Interposed between each cylinder and combustion chamber was a compressible copper sealing ring. A rectangular outer gasket fitted over the studs for water sealing." Hans Tanner and Doug Nye, *Ferrari*, p. 564.

[185] The 250 MM and Vespa scooter stand by while 0046M is being reassembled. *Audrey Saffell Sundin*

[186] The finished project with trees still bare of leaves. The third wiper post is gone and the superleggera script is back on the hood. *Audrey Saffell Sundin*

because of the lack of compression. That's the only one [engine] *I remember him pulling out. I know for a fact that it needed pistons besides rings and we couldn't afford it.*

Chuck reinstated the metal *Superleggera* script emblems on each side of the hood that Zagato had removed, and where he got them I do not know, but he did not return the Touring insignia badges that were originally above them. He also added a Sports Car Club of America emblem below the fuel filler cap where once had been the Zagato name, but he was not aware of that heritage **[187].** He removed the safety catches on each side of the trailing edge of the hood and replaced them with brackets on the hood and cowl for leather straps and buckles. Apparently, the straps and buckles were never installed because the brackets were mounted too close together **[188].**

He finally removed the third windshield wiper post that had gone unused for so long and filled in the hole but kept the two windshield washer nozzles. Prancing horse decals on each side forward of the doors and above the waistline completed the decorations **[189-190].**[179]

As Audrey indicated, the car was running again by the winter of 1957-1958 since the trees are still bare of leaves **[189-191]**. After surviving the winter under Audrey's blankets, 0046M was ready for a road trip a year late to Sebring, Florida, for the 12 Hours of Sebring race on March 22, 1958.The 1957 Michigan license plate was still on the car, however, and new registration was needed for the journey. Then, it would be a drive of about 700 miles (1,127 km) to Sebring,

[179] Figures 189 and 190 are from the estate of Jessie Coleman who was the perennial and colorful starter for sports car races all over the South in the '50s and '60s.

[187] Chuck has added a Sports Car Club of America badge where once had been the Zagato name. Straps hold the spare tire in place.
Audrey Saffell Sundin

[LEFT: 188] New brackets for leather buckles on the hood. *John Apen (Charles Saffell)*

[BELOW & OPPOSITE TOP: 189-190] Ferrari shield decals have been applied. *Jessie Coleman*

[191] The Michigan license plate is still on the car. *Audrey Saffell Sundin*

[192-193] 0046M parked at the Saffells' motel in Sebring and admired by bystanders. *Fred Leydorf*

an endurance contest in itself with two people in the cramped Ferrari with their necessities. As Audrey put it,

> *As you know, in that car you have no trunk space and that's how I learned to travel very light. You sure had to travel light in that car.*

After they had arrived in Sebring and were settled in their motel, probably tired and napping, the Ferrari was attracting attention parked outside. As I mentioned in Chapter 3, a friend of Fred Leydorf took photographs of the Ferrari parked and being admired in a classic '50s Florida setting **[192-194]**. Now the car bears the 1958 Alabama license plate 47-27921, indicating it has at last been registered in Huntsville's Madison County. The Scuderia Ferrari shield is missing, at least on the right side, perhaps a victim of poor adhesion.

They were well known in Southern racing circles by then, and according to Audrey, Chuck was one of the race stewards while she worked in timing. They had brought with them their copy of the *1957 Ferrari Yearbook*, no doubt proud of their inclusion, and filled it with autographs from famous drivers and notables. Many of those were probably collected at the awards ceremony where Audrey is seen in a photograph appearing in *American Racing: Road Racing in the 50s and 60s.*[180] Audrey identified herself standing at the upper right in dark slacks and blouse, holding a purse, behind a Who's Who of race drivers sprawled on the grass during the ceremony **[195]**. Romolo Tavoni is seated on the grass wearing glasses, tie, dark jacket and light-colored pants.

Since neither Chuck nor Audrey mentioned any issues, the car must have made the 1,400-plus mile (2,253 km) round trip without incident. The long trip probably entailed night driving, which might have been a reason why Chuck had added a pair of Lucas driving lights on sturdy brackets made of shaped angle iron and attached to the radiator mounts. (Those lights might have been retained from the traded TR2, which had a pair.) They were also making trips to Birmingham for meetings of the SCCA Alabama Chapter and those were held at night, leaving them a long return trip in the dark.

No doubt 0046M caught the eye of other SCCA members,

[180] Tom Burnside, p. 231.

[194] A 1958 Alabama license plate is now on the car. *Fred Leydorf*

[195] Audrey at the Sebring awards presentation – standing upper right in dark slacks and blouse, holding a purse. *Revs Institute*

and one proffered an offer to buy the Ferrari, which was accepted. Neither Chuck nor Audrey could recall exactly when they sold the car to Maurice Blevins, but Chuck did remember his wooden leg and mentioned that:

> *He apparently took it to a drag race the first weekend and blew the engine. That's the last I ever heard of the car. Ah, somebody told me he had some kind of an American V8 put in it.*

Of course I knew the engine had dropped a valve for some reason, but knowing Chuck's tendency toward colorful tales, I was skeptical about how much credence to put in the notion of a man with a wooden leg buying the car, much less racing it. He went on to say,

> *Blevins had a lot of money but not too many brains. I shouldn't say that probably, but car-wise he didn't.*

The wooden-legged Maurice Blevins became the missing link in the ownership chain of 0046M, and as I have said, I searched for him or any information about him for nearly twenty years with no results. Then, in 2007 I invited Charles Deerman to come see the old Ferrari I had bought from him. Deerman brought with him his bill of sale for the car, and when I read it I understood why I had never been able to find Blevins – his name was Maurice K. *Bivens*. That redirected my search, but it was still fruitless. By then internet searches were able to find anyone and they turned up a Maurice Bivens at an address in Chiefland, Florida, but he was not there.

Periodically, I would run a search for "Maurice K. Bivens" and finally after ten more years, in March of 2017, I found a 1968 Fort Walton Beach, Florida, newspaper notice that Maurice K. Bivens and Barney B. Burks, Jr., were issuing stock in their tire business. Fortunately, there was also a Florida voter registration record online for Burks, and it even listed his telephone number in Pensacola, so I called. After asking if he was the Barney Burks who was in business with Maurice Bivens back in 1968, which he was, I asked him if that Bivens owned a Ferrari in the early '60s. He said that he did and that Bivens was crazy about the car, but that he was now dead. I told him I had owned it and was researching its history, whereupon he said, "Oh, you need to talk to my sister, Geraldine – she was married to him."

Burks arranged a call for me with his sister and in our conversation the blanks were largely filled in. Naturally, the first thing I wanted to know from Geraldine was if Bivens had a "wooden leg," but I tactfully asked if he had a prosthesis. Indeed, she said that he did:

> *You would never know that he had a prosthesis. He could do anything. He just tried anything that anybody else could do that could walk. I don't think you would even notice he had an artificial leg. It was a birth thing that caused him to have it. I mean it was all his life so he was really agile. It was right below the knee.*

I learned that Bivens' middle name was Kevion and he was usually called "Kev." He and Geraldine divorced in 1971 and he had died in 1991 at 57 in Chiefland, Florida, so I would likely have missed him even if I had known his correct name. They were a young couple, married in 1955, and in 1958 lived near downtown Birmingham at 1611 9th Avenue South, Apartment 7. They both worked and Bivens was a draftsman for the Jefferson County Planning and Zoning Commission. He was a member of the Sports Car Club of America, which I confirmed from the 1958 Membership Directory, and that year she was 21 and he was 24. The Saffells were also listed in that directory and Geraldine remembered "Chuck and Audrey" from the club.

At that time, ownership of a sports car was a requirement for membership in the SCCA and Geraldine said they owned a Jaguar roadster, which they bought from the long gone Graffeo Motors in Birmingham, an XK140. Since Bivens is not listed in the 1956 SCCA Membership Directory but is in 1958, the Jaguar was probably purchased in 1957.[181] In her words,

> *We bought a brand new Jaguar roadster – it was a sporty little deal. I wasn't for it but he wanted to buy that Ferrari so we sold our Jaguar we were making payments on. And to me when we got the Ferrari it*

[181] I have not found a 1957 *Directory*.

> *just wasn't my cup of tea. I was never happy with it. I thought it was such a hunk of junk after having a brand new Jaguar, for goodness sake. But I didn't argue with him about it – he was so bent on buying that darned thing. When we had the Jaguar we were going to sports car events, but with that smoking, loud Ferrari I don't think he stayed with the sports car bunch then. It was just so loud and, oh God, and hey when I'm helping to pay for it and we had a beautiful car and then he got this urge to have that Ferrari. I didn't argue with him. I didn't even like to ride in it. We had another car. I would have divorced him earlier* [if she didn't have another car to drive]. *He had joined the sports car club and so I think that's where he got the idea about the Ferrari. He was just one of those guys who could never be satisfied with anything – including a wife.*

In Bivens' modest circumstances, the Jaguar must have figured into his purchase of 0046M from the Saffells, either sold for cash or used in a trade. Reconsidering what she had said earlier, Geraldine continued,

> *I don't remember selling the Jaguar. We had a discussion because that was the only way he could get the Ferrari was to trade in the Jaguar. Seems to me like – I don't know – well we would have to paid off the Jaguar...*

Bivens could have found a buyer for the Jaguar, or Graffeo Motors might have given him some equity compensation, or he could have struck some kind of deal with the Saffells to take it in trade. Mike Carpenter noted in our conversation that Chuck Saffell was known to be quite a "wheeler-dealer."

In a sequence of photographs apparently taken by Chuck, the two Saffell Ferraris were posed on their property side by side. In the first photograph **[196]**, Audrey is behind 0046M on the right. Geraldine confirmed that she is the young woman standing beside Audrey and Maurice Bivens is on the left. The exact circumstance of that day is lost to Geraldine's memory, but it is plain to see that Audrey seems to be speaking to her in earnest while Geraldine looks at the Ferrari and stands with arms crossed, her body language clearly communicating an obstinate disapproval. Meanwhile, the candid photograph caught Maurice in what appears to be a gesture of frustration. In the second photograph **[197]**, of better quality, the Bivenses' white Jaguar roadster is visible in the background while Audrey is posed in the 250 MM and their son, Arthur, is standing by 0046M. There are several other photographs in this series, including **[198]** and **[199]**, and it could be that they were taken as farewell mementos on the day of the sale, with the Jaguar to be left behind in trade, or they could record a preliminary visit to convince Geraldine. In these photographs, the grass is

[196] Audrey (right) seems to be reassuring an obstinate Geraldine Bivens about 0046M while an exasperated Maurice Bivens stands by. *John Apen (Charles Saffell)*

[197] Audrey sits in their 250 MM while son Arthur stands by 0046M. The Bivenses' XK140 is in the background. *Audrey Saffell Sundin*

[198] Audrey poses by 0046M. *Audrey Saffell Sundin*

[199] The 1958 license plate on 0046M as it sits by the 250 MM. *John Apen (Charles Saffell)*

green, the trees are in full foliage, and it must be summer because Audrey and Arthur are in shorts. Arthur has grown and would be three-and-a-half years old in the summer of 1958, to which his size would attest.

The Saffells, though, had become Ferrari people, and if they did take the Jaguar in trade, it would not suffice. Soon after the sale of 0046M they acquired a 250 GT Boano Ferrari (s/n unknown) with 5,000 miles and could possibly have used the Jaguar as a trade on it. In **[200]** 0046M is now no longer in the picture, and the 250 MM and the 250 GT have been posed side by side for a photograph in their driving suits that appeared in the 1959 *Ferrari Yearbook*. Figure **[201]** is a different view of the scene and Arthur is beside them. He has not noticeably grown since he was photographed beside 0046M and even appears to be wearing the same clothes **[197]**, so no appreciable time has passed. At a time that was probably not long after this photograph was taken, they sold the 250 MM as well. It seems Chuck entered the car in the Courtland race for August 31 but it did not show up; then, it was raced by its new owner elsewhere on September 7.[182] That places the sale of the 250 MM probably in August and the sale of 0046M to Bivens before that, maybe June or July. The fact that their photograph is in the 1959 *Yearbook* is irrelevant for dating since the photograph could well have been taken in 1958, and surely was.

[200] Chuck and Audrey pose for a photograph with their Ferraris for the 1959 Ferrari Yearbook. 0046M has been sold and a 250 GT has taken its place. *Ferrari S.p.A.*

[201] Son Arthur joins Chuck and Audrey for a photograph with the two Ferraris. He wears the same clothes as during the Bivenses' visit. *Audrey Saffell Sundin*

Chuck and Audrey continued racing with other cars through 1959 and then divorced in January 1962. Eventually, Audrey returned to Southern California, while Chuck eventually moved to Florida and opened Competition Motors.

Bob Rabun, from whom I bought the engine, knew Bivens and 0046M from the Birmingham meetings of the SCCA and they lived perhaps a mile (1.6 km) apart. According to Rabun,

> *I remember meeting Blevins* [sic] *at a sports car club meeting. Ok, and then it seemed like he didn't come to the next one and I didn't see anything of him for about six months and then I did see him again and I asked him how he was doing and he said, well, he had been rebuilding his engine, in that Ferrari. And in fact one or two months after that he drove it to a meeting. I do know that he blew the thing up twice and when I bought it was after it had been blown up the second time, not the first time. Now, he did not know enough about engines to do that himself. I know beyond a shadow of a doubt that whoever he got to do it did not know anything about Ferrari engines and probably never worked on foreign car engines. In fact, he told me at the time who he got to do it and it was somebody in an independent garage and a small one at that that normally just worked on run-of-the-mill clunkers really. Which I mean that's tragic but that's the situation.*

[182] Currently, one source has the 250 MM (0348M) sold in late 1957, but since **[199]** shows it in summertime sitting beside 0046M with its 1958 license plate in the Saffell's yard, that cannot be correct.

Rabun ascribed Bivens' rebuild to his having blown the engine, but that was probably not the case because there was no internal evidence that any damage had occurred other than the incident in one cylinder that finally disabled the engine. As Geraldine complained, the car smoked, and that was because Chuck Saffell had not put in new pistons on his rebuild of the motor. The same sensitivity to cold starting that Audrey described in the winter of 1957 would have become apparent to Bivens when winter came in 1958. At that time, they were living in an apartment building in urban Birmingham, so there would have been no sheltered parking for the Ferrari nor the possibility of putting a heater under it as Audrey had done. Since the car was Bivens' means of transportation, he probably found himself stranded on many cold mornings and I suspect that led him to have the engine rebuilt to recover compression and relieve the smoking.

Geraldine did not have any recollection of the woes and work involved with the Ferrari, and she probably could have cared less. Her husband probably never realized what he was getting himself into, since Rabun summed him up this way:

> *He seemed like a misfit for the car because the car seemed fine and he seemed like, uh, a little bit less than Joe Blow, really. It didn't fit his personality or temperament or anything. He was married and no doubt didn't have a ton of money to put into the thing, and I'm gonna say that at the time he had to be younger than 27, possibly younger than 25.*

Bivens' choice of an unspecialized shop to work on the engine was likely dictated by economics, but the early Ferrari engines are particularly tricky and require some knowledge, not to mention the right parts. Rabun said,

> *When we disassembled the thing, every rod bearing and every main bearing in that thing had been Permatexed – he'd used that to sort of glue them into place.*

The engine did run after the rebuild, but eventually either malfeasance or fate brought about a catastrophe: a valve spring broke, the exhaust valve in the number ten cylinder hit the piston, and the proverbial irresistible force met the immovable object. I consider the drag racing story to be dubious and the breakage could have been caused by the failure to assess the condition of the springs and valve keepers or incorrect assembly, but it also could have been due to driver error. Hans Tanner noted, "The gearbox is somewhat of a problem until the driver has had a certain amount of practice with it, and although third and fourth gears are synchromesh, selection can be accompanied by anything except the silence claimed for it."[183] According to the Assembly Sheets of 0046M, however, its transmission had no synchronization, and John Bond described shifting gears in a 166 Mille Miglia this way:

> *Most people think a 5-speed crash gearbox... is difficult. It is, but then it isn't if you know how... But downshifting is another matter. Going from 5th to 4th to 3rd doesn't call for double-clutching, though it helps just to avoid a crunch. But 3rd to 2nd requires double-clutching, and 2nd to 1st – don't try it unless you're a Phil Hill.*[184]

In 1957, Paul O'Shea wrote about double-clutching in his book *Guide to Competition Driving*:

> *Once upon a time, shifting a gearbox was quite a trick in a sports car. High revving engines made the process not too smooth going up and pretty rough going down. Usually, the* [down] *shift required double-clutching – that is, depressing the clutch pedal, pulling the stick into neutral, letting up on the pedal, giving the accelerator a burst, depressing the pedal again, pulling the stick into the chosen gear, letting up on the pedal... Formerly, therefore, the man cultivating his driving prowess had to devote long and arduous practice just to shifting.*[185]

Audrey Saffell would have objected to O'Shea's sexist assumption that the driver would be a man, but in other words, double-clutching requires a lot of artful and dexterous coordination between the shifting hand – left in this case –

[183] Tanner, p.12.
[184] *Road & Track*, February 1973, p. 65.
[185] pp. 39-40.

and feet on both clutch and gas pedal. Regardless of how adept Bivens was with the prosthesis on his left leg, he might have miscued during the complex left footwork on the clutch during a downshift, left the transmission in neutral and over-revved the engine, thinking he was in gear. This mistake would be easy enough for someone with a fully functional left leg and the result could well be a broken valve spring and dropped valve. That is what I suspect happened.

Faced with the insurmountable expense of repairing the blown Ferrari engine, Bivens turned to the expedient and cost effective solution that many Ferrari owners opted for in the late '50s and '60s: replace the engine with the compact and more powerful Chevrolet 283 cubic inch V8. According to my departed friend Randy Cook, there was an article in the March 1959 issue of *Sports Cars Illustrated* that explained in detail how to swap a Chevy V8 into a Ferrari 212, which has a chassis similar to the 166, and in July 1959 *Motor Trend* did the same with a different Ferrari.[186] So, conveniently the blueprint was available for the conversion to whatever shop it was that did the work for Bivens and those articles would have been published before the engine swap. Rabun gave me a perspective on the sequence of events:

> *He kept the car for a while; in fact he kept it with the Ferrari engine something like two years, then seems like he had it with the Chevrolet engine for about two years or so after that.*

Two years with the Ferrari engine would place the swap in the last half of 1960. Rabun speculated that it was in November 1961 that he and Mike Carpenter bought the Ferrari engine, and Carpenter agreed. By that time, the Bivenses had moved from their apartment building downtown to a separate apartment behind a house in a suburban neighborhood, 830½ South 78th St., and that was where Rabun and Carpenter collected the engine.

Suburban living proved to be a challenge for someone with a race-bred Ferrari, though. There was still no garage and the car could only be parked in the driveway or on the street. Geraldine lamented:

> *You know about the big tank for gas and the big gas cap. We had a problem. At the time we lived in a little old apartment in Birmingham; we didn't have a garage. Kids put sand in there one time and they put rocks and put sugar in with the gas. They opened up the cap and put that in there.*[187] *We didn't have a garage so it was just sitting out on the street. You'd have to sleep in it to keep the kids out.* [There were no locks on the doors.] *It got to be more of a problem than a pleasure; of course it never was a pleasure to me. It just got to where it was not feasible for us to keep it.*

Bivens' love for the Ferrari is quite apparent from all the effort and money he must have lavished on it. Eventually and most likely begrudging having to do so, he put his prize up for sale for $3,000,[188] as advertised in the *Birmingham News* on Friday, June 15, 1962 **[202]**. The "new interior" he claimed could have amounted to carpeting, since with the engine swap there was a new transmission and shifter to cover, but it is hard to say if the seats and headliner were redone or not. Although the car had been repainted in 1957, Bivens' claim of new paint seems to be true.

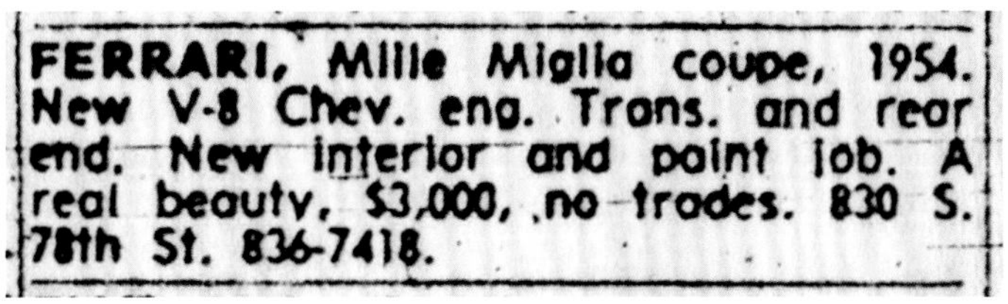

FERRARI, Mille Miglia coupe, 1954. New V-8 Chev. eng. Trans. and rear end. New interior and paint job. A real beauty, $3,000, no trades. 830 S. 78th St. 836-7418.

[202] Bivens' advertisement in the *Birmingham News*, June 15, 1962. *Birmingham News*

The photographs in **[203-206]** were taken by Gerald Roush in response to the advertisement and show 0046M as presented for sale with its Alabama license plate 1-93934. The Bivenses' apartment was the building at the end of

[186] *Bowtie Ferraris*, p. 9.

[187] Fortunately, there is a filter that inserts into the bottom of the tank through a threaded opening to which the fuel line is connected at a right angle, so draining and flushing it would not have been too difficult.

[188] In 2024 dollars: $31,340.

[THIS PAGE AND FOLLOWING: 203-206] Gerald Roush's photographs of 0046M taken when he responded to the advertisement. Bivens has repainted the Ferrari and the windshield washer nozzles have been removed. The Bivenses lived in the building at the end of the driveway. *Gerald Roush*

HEART OF DIXIE
1 -93934
ALABAMA 62

the driveway **[206]**. The windshield washer nozzles are gone and the holes have been filled in. Furthermore, Chuck Saffell had left the engine compartment painted dark gray in **[184]** (see page 185), but when I bought the car, the engine compartment had red paint, as seen later in **[247]** (page 230). The opportune time to paint it was before the Chevy V8 was installed and the whole car must have been re-sprayed red then.

Charles Deerman of nearby Tuscaloosa saw the advertisement and came to see the car. Deerman was 27 years old and had a budding interest in sports cars, so he offered Bivens half of his asking price and made a deal at $1500 for the Ferrari.[189] Mirroring Bivens, Deerman already owned a Jaguar, a 1952 XK120 roadster, and throughout his life he accumulated quite a store of interesting sports cars.

Back in Tuscaloosa with the Ferrari, Deerman dated his future wife, Ruth, in it. Unlike Geraldine Bivens, Ruth was charmed with the car and they were married a month later in July. She drove the car herself on occasions and told me:

> *Every time we would drive it, people would, you know, stop us and ask us about it. We got a lot of attention. We thoroughly enjoyed it when we had it.*

[189] In 2024 dollars: $15,670.

It was an uneventful four years that the Deermans had the car, using it for pleasure since they had other vehicles, and in 1966 it carried the Alabama license plate 63-27562 for Tuscaloosa County. Now and then, though, Deerman would drive the Ferrari to work from their home on 14th St. East in east Tuscaloosa to Hunt Oil Company on the west side of town. They lived only a stone's throw off University Boulevard, which traversed most of the distance, and taking that route carried him through the University of Alabama campus. It was on one of those random trips through campus with 0046M that fate brought us together in the spring of 1965.

Roughly a year later, in 1966, Deerman decided to repaint the Ferrari once again to "spruce it up," as Ruth put it, perhaps impugning the quality of Bivens' previous paint work. According to her, he did the paint removal himself having "bought the equipment" – a sandblaster – to speed up the work. The resulting removal of so much filler in the process likely prompted him to hand the car over to someone more expert in bodywork. Ruth was a native of Columbus, Mississippi, and through connections he engaged someone there to finish the project, leaving the Ferrari with them. Before any work began, however, I showed up on his doorstep on June 29 offering to buy it.

Chapter 10

1966-2007: MY FORTY YEARS

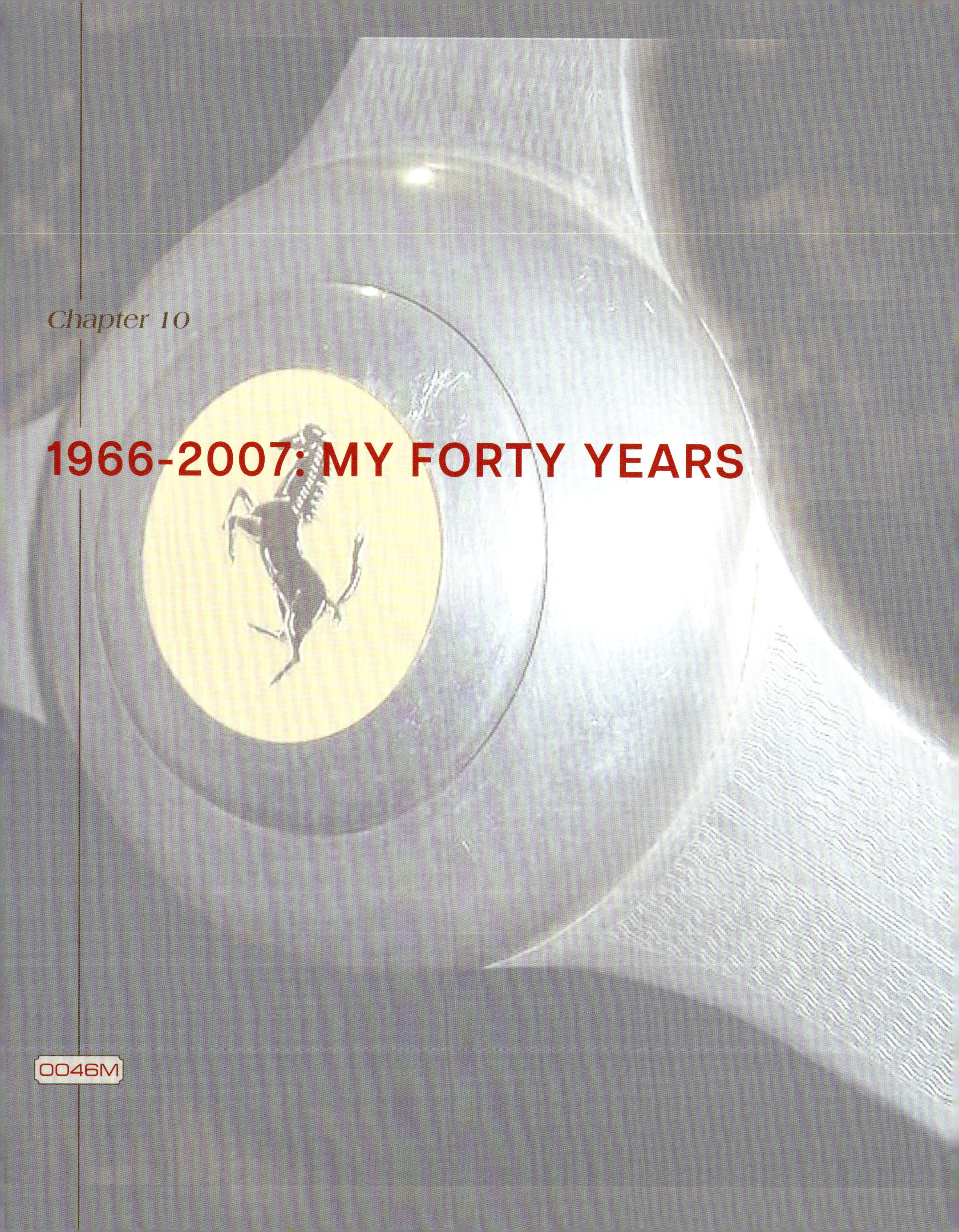

I have already related how I bought 0046M from Charles Deerman on July 5, 1966, and left my part of the story with the car stored in a rickety little garage behind my mother's house in Mobile, Alabama. Eventually, I decided the car really needed to be sheltered in a safer place than the old garage. By then I was established and had a huge, secure warehouse and shop for my business where I was already storing my E-Type Jaguar and a Mercedes 190 SL I had bought, so it was time for the Ferrari to make a return trip to Tuscaloosa. In late October 1988, I borrowed a car trailer from Frank Wilson, who had a local restoration shop, and had a hitch put on the 1986 Ford Ranger I was driving at the time.

The sequence of photographs in **[207-212]** shows the Ferrari in the garage and the removal and loading of the car on the trailer. Complicating the extraction, the floorboards under the wheels had given way over time and the car was now below the threshold of the door. In the last photograph, I am sitting on the trailer resting after the job is done with the engine and other parts in boxes in the bed of the truck. All that remained was a 200-mile (322 km) drive with the overloaded and underpowered 3-liter Ranger. But, with

[THIS PAGE: 207 & 208] The car is covered with years of dust, dirt and pigeon droppings. The floor of the old garage at my mother's house has collapsed under the Ferrari, making removal difficult. *Author*

[ABOVE: 209] Extraction involves pulling the car out with my truck in the cramped space of the back yard. *Author*

[LEFT: 210] My mother looks on as the moving progresses. *Author*

[211] Once at the trailer, the winch takes over. *Author*

[212] With 0046M on the trailer and the engine parts in the bed of my truck, I take a break. *Author*

perseverance, I made it to my warehouse and was proud of myself when I backed the long, wide trailer through the overhead door, threading the needle on the first try.

The next morning, I had my employees unload the car from the trailer and let them satisfy their curiosity before the day's work began. They gave it a bath to remove the layer of dirt and pigeon droppings that had accumulated in the old garage and then we moved it to a back corner where it would be safely out of the way **[213-215]**. There it spent the next 19 years, joined over time by other cars in my collection.

I had no intention of parting with my Ferrari, regardless of how much I saw that it was appreciating in value. My long-term goal – although I had no idea how long the term might be – was to restore it, re-install the Ferrari engine, and surprise the Ferrari community with a car that had been "missing" for decades, and to boot I would have documented its full history. In spite of that, it was not as though I was hiding the car. I had consulted a number of historians and all of my acquaintances in the Ferrari Club of America knew about it. As you might expect, I had inquiries from time to time, calls from brokers who asked what I would take for it. My stock reply was 1.5 million dollars, which I knew was a bit much for a fully restored 166 MM *barchetta* at the time, so that would end the call.

As prices for rare old Ferraris continued to move skyward through the '90s, a treasure hunt mentality arose in which people were ferreting out cars worn and weary from their lives in domesticity, buying them for what their owners regarded as a windfall, making some cosmetic and mechanical improvements, and then flipping them to new and wealthy owners who quite likely would wait until prices rose further and then sell them again, like stocks. There were even investment groups formed to acquire the most desirable – and costly – Ferraris, expecting a jackpot. Of course, none of them drove the cars on a beer run, or sat in them, or knew the smell of an old car. They became commodities, but I wanted to keep my Ferrari what it had always been: a *car*. I came to think of it as one of the last early Ferraris not in captivity.

Since I had accomplished much of what I had set out to do in learning the history of 0046M, my research slowed as new details became harder to find, but I would pick up the

[213] Ferrari and trailer, safe in my warehouse. *Author*

[214-215] 0046M's new home in a back corner of my warehouse. *214: Author, 215: Winston Goodfellow*

chase from time to time. Over the years, I wasted a lot of effort trying to find the non-existent Maurice Blevins, but I did have other successes.

All along, I looked for things that would be needed for the restoration. I met Sherman Wolf, another 166 MM owner in those days, at the 1989 Ferrari Club of America National Meet in Atlanta, and in talking about the things I had missing, he offered to look for electrical parts on a future trip to Italy. A year passed, but eventually he was true to his word and sent me a Magneti Marelli key switch, a starter pushbutton with center light, and a left-off-right turn indicator switch, all period correct and scarce.

Although I had a correct transmission, albeit with the wrong shifter, and I had the original driveshaft tunnel, I did not have the original transmission cover and had no idea how it should be shaped. That problem was solved in early 1990 when Scott Restorations made a copy of one for me in appreciation for my help in their reconstruction of a firewall for the restoration of 0054M. The cover turned out to be a complex thing and was a beautiful piece of workmanship.

I wanted to learn more about the races my Ferrari had run and I knew that one of the major Italian motoring publications in that era was *Auto Italiana*, so I advertised in *Ferrari Market Letter* wanting to know if anyone had a collection of the magazine covering 1950-1954. I got a call from Scott Borman in Los Angeles who said he had many of the issues from that period, but he had no interest in selling them. He was agreeable to making copies, though, but to Xerox them would be bulky and costly. After some thought and phone calls, I arranged to have the relevant issues microfilmed by a library reproduction service in Los Angeles in January 1991, and once I had the reels in hand I was able to study the magazine at leisure in my local library, finding much valuable information. There were several key issues missing, unfortunately, but I learned that the Library of Congress had a complete set of *Auto Italiana*. So, when the Ferrari Club of America held its annual meet in Virginia outside Washington, D.C., later in 1991, I went with the additional purpose of spending a day at the library, where I was able to find and copy the missing material. It was through those two efforts that I was able to learn many of the race details given in the earlier chapters.

In 1998, I had a chance to inspect another 166 MM, 0058M, undergoing restoration at Bob Smith's Coachworks in Gainesville, Texas. Since I had the transmission and differential but no driveshaft to go between them, I wanted to get detailed measurements and photographs of an original in order to fabricate one for my car. Securing an invitation, I flew out to Dallas, drove a rental car the 70 miles (113 km) up to Gainesville, and began inspecting the 166 which had been totally dismantled; even the body had been removed from the framework it tightly enveloped. Everything was familiar since I had crawled in, under and all around mine. Furthermore, this was formerly the car of Ed and Sherry Gilbertson that I had looked over at the *barchetta* reunion. Since the frame was bare, I had easy access to the engine mounts, which had been cut off on my car and replaced for the Chevy V8. I took measurements and photographs and made cardboard templates for future reference. Likewise, I photographed and measured the driveshaft and the rear axle assembly, since I was missing the right side axle housing. It was an amazing opportunity to see a 166 MM restoration in progress and I spoke with several of the craftsmen. They were happy to meet me since, as one said, the owners of such cars as they worked on never came by to talk to them.

Having examined the chassis of 0058M, I went looking for my chassis number when I returned. There was such certainty that I had 0046M that I had never bothered to find the number stamped on the chassis, and actually before the Texas trip I did not know where it was. Armed with a flashlight, I leaned over the left fender at the wheel, looked down into the engine compartment and found the mount of the left front Houdaille shock absorber. The number should have been stamped on the frame tube directly behind that but what I saw alarmed me. The area was covered with years of road grime, but much worse was that near where the number should be was a motor mount welded in for the Chevy V8 over the cut-off Ferrari motor mount. If the number had been destroyed, that would have been a serious detriment to originality. I scraped away layers of grime between the shock and the motor mount, getting down to bare metal, and with great relief I found the number. It had barely survived, though, since the welder had ground the area with a disc sander in his preparations and had nearly obliterated the 4, 6 and M, plus welding slag had fused to the number. In the right light it was legible, fortunately, and I was grateful for that **[216]**.

[216] The chassis number on the frame, slightly damaged during the welding of mounts for the Chevy engine. *Author*

And then there was a mysterious incident that adds to the lore of 0046M. Before I brought the car back to Tuscaloosa, I had shown photographs of it to my employees. One of the guys liked to hang out at Harry's Bar and one day he told me that there was a picture of my Ferrari hanging on the wall there. I skeptically disregarded that news, since to the untrained eye many old Ferraris look alike. Many years later, I got a call from Harry. He needed a coin acceptor for his cigarette machine and offered to trade me the picture of my Ferrari for one. I decided it was about time to see what this was all about, so I carried the part over to the bar. Harry led me to a cluttered back storage room and unearthed a nicotine-caked, 2 by 3 foot (61 cm x 91 cm) framed picture. I was stunned. It was a painting of 0046M **[217]**. Harry said a guy had come into the bar long ago and offered to trade the picture for a few beers, so he added it to his college bar décor and never saw him again. Fortunately, there was a

[217] A painting of 0046M done by Duncan Black of Montgomery, Alabama, in 1958 and found at Harry's Bar, Tuscaloosa, in the 1990s. *Author*

sheet of plastic protecting the painting from the elements in the bar, and when I removed that it was in good shape. On the back I found an inscription: "Duncan Black 1958." The SCCA Directory for 1958 lists Duncan Black as a member living in Montgomery, Alabama. Most likely Black had seen the car and was impressed enough to commemorate it with his painting – in a whimsical Mille Miglia setting. That might well have been at the 1958 Courtland, Alabama, races on August 31 when the Bivenses could have had the car there, but where was it for at least 20 years before it turned up in Harry's Bar? Now the painting hangs on the wall over my desk.

Years slipped by with the Ferrari patiently waiting in the back corner of my warehouse. As 2005 got underway, the amusement business was waning and lacking the means and the ability I began to come to terms with the likelihood that I would never be able to restore the car. Reluctantly, I began to entertain the thought of selling it so that it could receive the attention that it needed, if I could find a good home. One day I received a call from Doug Freedman, a friend from the past when I was active in the Ferrari Club of America. Doug was highly active and well known in Ferrari and classic car circles as a collector, as a Concours judge, and as the organizer of the annual Concours on the Avenue in Carmel. Doug said that he had mentioned my car to David Sydorick, who specialized in collecting Zagato cars, and that he had shown interest. So, Doug wanted to know if I would like to talk with him about the possibility of a sale, to which I agreed, thinking it would be a good way to judge the market potential.

Sydorick and I spoke by telephone and then he made the trip from his 90210 zip code home in Beverly Hills to Tuscaloosa. Over dinner I gave him a condensed version of 0046M's history and the next morning, a Saturday, I brought him to my warehouse to see the car. He was interested but wanted a professional opinion, so after he returned to California he sent Wayne Obry of the prestigious restoration company Motion Products to inspect the car. Obry looked it over thoroughly for his report to Sydorick, and before he left I asked him what he thought a restoration would cost, which he estimated to be $300,000.[190] That piece of information solidified my decision to sell, as I realized I would never be able to afford a proper restoration. In a few days, Sydorick called again and offered much less than I felt the car was worth, so I declined and that episode ended.

As well as learning the restoration cost, I gained another insight from the Sydorick interlude. When I did internet research on him, I came across an article on his Aston Martin Zagato in which the prior owner explained his handling of the sale. He had created a charitable remainder trust with Cal Tech University, donating the car to them, and the proceeds from the sale went into the trust, effectively sidestepping the 28% capital gains tax on the sale of collectibles and allowing the full amount of the proceeds from the sale to be invested and grow. The seller received a fixed percentage of the trust's worth each year and the capital gains tax was only applied against that dispersal. The catch was that the IRS regulations allowed only two successive beneficiaries of the trust and upon their deaths the remaining trust amount would revert to the designated charity. I knew about the 28% tax on collectibles and had been concerned about losing so much money, since my appreciation would be enormous, so a charitable remainder trust seemed the perfect answer for me. My wife and I had no children, so the income would benefit us through our lives, first accruing to me and then to her. Now I had a plan.

My next step was going to the Office of Advancement at the University of Alabama, which is charged with fundraising, principally through donations. I asked to speak to someone about a donation and was introduced to Phillip Adcock, Assistant Vice President for Development. In his office, I began to explain that I had a car I wanted to donate to the university through a charitable remainder trust and that I had a precedent I could show him that had been done with Cal Tech. Recalling the conversation later in an interview,

[190] In 2024 dollars: $484,630. During his visit, Obry told me that his company had restored the 250 MM that the Saffells had owned and he said that they had to research and replicate one part: the missing passenger side windscreen for the car. Hearing that, I opened a nearby box and showed him the original that Chuck had sent me as a gift.

Adcock said,

> *We hadn't worked with Mr. Westfaul before... so when he told us in our initial meeting that he wanted to fund a unitrust with a used car, I was a bit apprehensive. But I was patient and just listened.*[191]

He thought at first I was talking about an old Ford or Chevrolet, but when I told him it was a Ferrari and I thought it should be worth well over a million dollars, I got his attention. Needless to say, the university was happy to work with me and we agreed that when a sale looked imminent we would begin the formation of a trust.

When he heard that Sydorick was out of the picture, Doug Freedman called me back to say that he would soon be in Europe and would be at a dinner with Pierre Bardinon, the owner of a once fabulous collection of Ferraris, and would see if there was any interest there. Bardinon, it turned out, was going in the other direction, downsizing the collection, but word went out that 0046M was for sale.

At some point in 2006, I called Wayne Obry, since his customers were well-heeled collectors, and asked him to check with people he knew to see if they had any interest in the Ferrari. However, no one ever contacted me through the overtures of either Freedman or Obry.

As 2006 drew to a close, conditions in the coin-op amusement business had seriously deteriorated. Computer games had already killed the video game business, while the growing recession had reduced discretionary income and now playing other coin-op games was deemed an unnecessary expense. In addition, it would not be long before I would have to make major capital reinvestments in equipment and vehicles. I had been running my business around 32 years by then, and I was 62, so it was time to consider retiring. To that end, I quietly began making plans to sell my business, but it was plain that it would take the sale of my Ferrari as well to achieve a good retirement income.

On the evening of January 14, 2007, I logged on to the website *ferrarichat.com* under my pseudonym and checked the vintage Ferrari forum, since I had not been there in some time. There I saw a post that had been made the past December 25th:

> *I am hunting for a 166 MM to buy. Does anyone know of one?*

Deciding to take the bait and see where that would lead, I replied privately:

> *If you seriously want to buy a 166 MM, please send me a return message with contact info. I own one.*

I did get a reply by private message and that began a string of correspondence with Dean Micalizio. He turned out to be an amateur broker looking to play in the high stakes game of Who's Got the Ferrari, but he claimed to have a buyer for a 166 Mille Miglia whom he was representing. That buyer, he said, was looking for an original condition *barchetta*, so he would not want my car, but he felt sure he could find a buyer for me and wanted to know my price.

All along I had been guessing at the figure of $1.5 million as the car's worth, but I realized that since no one had shown interest so far, I should take a realistic look at what its value might be. 166 Mille Miglias had rarely come to market and mostly had changed hands by private sale, but 0054M had sold at auction for $1,760,000 in 2005.[192] That car was a *barchetta*, however, and it had been fully restored, whereas 0046M was no longer original, even if it was unique. Furthermore, my engine was dismantled and in dubious condition and the car was too. History was a trade-off, since most 166 MMs had racing history. What I decided as a fair price would be a half million or so less than what the original, restored car sold for, to wit $1,200,000.[193]

It was not long before Micalizio found an interested party but his resources were not such that he could pay for the car outright, so he kept offering alternative ways to finance

[191] www.sharpenet.com/give-take/commitment-education-drives-gifts/.

[192] In 2024 dollars: $2,826,900.

[193] In 2024 dollars: $1,815,500.

it and we dickered with him for a couple of months before abandoning that prospect.

While those fruitless negotiations were going on, I prepared on my end by setting up the charitable remainder trust and Adcock put the university's legal department to work on the details, creating what is known as a "flip unitrust." After lengthy delays and discussions, the agreement was sent to the Board of Trustees and they approved. I signed the trust agreement on April 11, 2007, and brought a model of 0046M to the signing **[218]**. The Ferrari then belonged to the trust but I was the trustee and it would remain in my possession until I sold it, which is known as a "triggering event," whereupon the flip would take place and the university would become the trustee with the proceeds of the sale going into the trust. The trust would then pay me or my wife a fixed percentage of its annual value until we were both deceased, at such time it would become a scholarship fund. I specified that the scholarships would benefit University of Alabama students from Murphy High School in Mobile where I had graduated.

Following the trust signing, I gave Micalizio an exclusive representation for 90 days and sat back to see what would happen. What I did not know was that he in turn would go to other, apparently more well-connected, brokers with his deal and become a middleman. Some names involved were Terry Godbout, Steve Yanoshek and Phil Handley, although I was never in communication with any of them until the last minute. It seems they, too, made little progress until finally another collector with an interest in Zagato came forward. I was told someone in Arizona who had a number of Zagato cars in his collection was forming a museum and

[218] I sign the agreement with the University of Alabama that put 0046M into the Darrell Westfaul Charitable Remainder Trust. Phillip Adcock, Assistant Vice President for Development, looks on. *Chip Cooper*

that he wanted my Ferrari. The secrecy and behind-the-scenes dealings bothered me because I wanted a personal involvement and wanted to be certain that 0046M would be going to a good home rather than become a piece of merchandise bounced from one high bidder to the next. All I could get were assurances that the buyer was reputable, that his intention was to restore the car and exhibit it in his museum and that my price would be met.

Even if the process was not what I had hoped for, at least it was a deal and that had been hard to come by, so I accepted it. The university seamlessly flipped the donated payment of $1,200,000 into the trust and took over as trustee when I sent the bill of sale, dated May 21, 2007. By then I had been made aware that the buyer was Scott Gauthier, a high-end jeweler in Scottsdale, Arizona. The brokers involved added their commissions to the purchase price and I have heard $175,000 as the figure they split among themselves.[194]

While the brokers wanted to cloak the transaction on their side, the university saw a publicity opportunity and the Office of Advancement sent out press releases about my "gift" to the school, hoping to inspire other donors. In the internet age, Warhol's 15 minutes of fame reverberates around the globe and resides in servers indefinitely. Google showed me news stories around the world repeating the line, "Businessman donates million dollar Ferrari to his alma mater," or something similar. The problem was that in repetition the stories rarely explained how the trust worked and it seemed as though I had made an outright gift rather than retaining a financial interest. The story was printed in my local Tuscaloosa paper and was front page news in the *Mobile Press-Register*. Friends, relatives and all kinds of people were convinced that I must be very wealthy, so one by one I had to disillusion them and explain what the situation really was.

Once I knew the Ferrari would be leaving, I called Charles Deerman, who still lived in Tuscaloosa, and invited him to visit his old car. Even though he was in bad health, he could not resist and his son-in-law brought him to my shop where he enjoyed the reunion and our visit **[219** and **220]**. It was then that he showed me his bill of sale and I learned that Blevins was really Bivens. Deerman passed away about 18 months later.

[194] In 2024 dollars, the total price was $2,080,250.

[219] Charles Deerman reunites with his old car. *Author*

[220] Deerman and I shake hands again after 40 years. *Author*

All that remained was for me to inventory and safely pack up all the parts and prepare the car to be shipped to Arizona. It was a sad time for me, handling all the pieces that I knew so well from 40 years of working with them and living with them. The Ferrari was like a member of my family and it felt as though I was losing a part of myself. It was the end of an era covering most of my life and no amount of money could have cheered me about that.

But, I wanted my car to be whole again and if I could not make that happen, then I wanted the new owner to have everything in order so that he could. So, I packed and labeled everything carefully, taking precautions to protect the fragile things, like the interior trim around the windows and the two pieces of glass from the windshield, especially the one with my University of Alabama parking decal. Appendix 18 is the packing list I prepared. I managed to fit everything into the car that belonged to it, except for the fuel tank, transmission, and engine parts, all of which had been crated for years. On the following pages, **[222-224]** show what was involved and how everything was fitted into the Ferrari, and then it was time for reflection on our life together **[225]**.

Finally, the day came for the transport company to pick up 0046M: June 11, 2007. We had rolled the car to the front of the shop the day before and when the transporter arrived, my men pushed it out and onto the lift gate, then on inside, followed by the extra crates. Soon, it was all buttoned up and I watched the transporter pull away and out of the gate, leaving behind forty years of memories **[226-230]**.

WEATHER

Today: Mostly cloudy. Highs upper 80s to around 90. Rain chance 40 percent today.

Tomorrow: Mostly sunny. Highs in the low 90s. Lows 60s and 70s. Rain chance 20 percent.

Complete Weather/8B

BAYWEEKEND
MUSICAL TREASURES DISPLAYED
PAGE 1D

TOMMY WILCOX OUTDOORS
EX-BAMA STAR HOSTS TV SHOW/1C

WAL-MART: DAMAGED STORE EXPECTED TO REOPEN SOON/1B

MOBILE EDITION
www.al.com

PRESS-REGISTER
SERVING THE ALABAMA GULF COAST SINCE 1813

THURSDAY
July 5, 2007
50 Cents

IRVINGTON

Press-Register photo

Organizers of the International Community Center are: front row, Phuong Phanly, left, and Sue Chathapaseuth; second row, Chanda Tan and Josh Vannavilaithong; and in the back, executive director Pat Sivoravong.

Office serves as 'hub' for aid

▸ Asian-American group opens International Community Center

By RUSS HENDERSON
Staff Reporter

IRVINGTON — A group of Asian-Americans on Monday opened the doors of a new office, a former three-bedroom rental home on U.S. 90 that members hope to transform into an "international hub" of information and post-disaster

'MORE PATIENCE, MORE COURAGE AND MORE SACRIFICE'

AP photo

With those words, President Bush implored the nation not to lose faith in his Iraq policy. In a patriotic Fourth of July speech Wednesday in Martinsburg, W.Va., Bush said, "If we were to quit Iraq before the job is done, the terrorists we are fighting would not declare victory and lay down their arms. They would follow us here."

COVERAGE/PAGE 6A

$1 MILLION FERRARI TO PAY SCHOLARSHIPS

BUSINESSMAN SELLS RARE CAR TO GIVE MURPHY STUDENTS THE OPPORTUNITY TO FOLLOW THE ROAD HE TOOK TO BAMA

Photos by CHIP COOPER/University of Alabama

Above: Darrell Westfaul stands with the 1950 Ferrari that he recently sold for more than $1 million. The money will fund University of Alabama scholarships, for which Murphy High School graduates will be given first consideration. **Below:** A university parking sticker from Westfaul's days at Bama remains affixed to the Ferrari's windshield.

Sugar, Splenda in legal fracas

▸ Trade group seeks court order to inspect sucralose plant in McIntosh

By BRENDAN KIRBY
Staff Reporter

Lawyers for a sugar trade group that alleges unfair advertising practices by the marketers of Splenda have asked for a court order to inspect the plant in McIntosh that makes the artificial sweetener's key ingredient.

U.S. Magistrate Judge William Cassady on Tuesday scheduled an August hearing on the matter in Mobile's federal court.

The request comes as part of a lawsuit that pits The Sugar Association against McNeil-PPC and McNeil Nutrionals LLC. The lawsuit, scheduled for trial in January in a California federal court, accuses McNeil of misleading advertising. The association takes issue with the advertising

London-based Tate & Lyle, which owns the McIntosh chemical plant that makes the sucralose used in Splenda and employs about 160 people, has refused requests to allow The Sugar Association to examine and videotape the manufacturing process.

[221] My donation of a million-dollar Ferrari is front page news in Mobile. *Mobile Press-Register*

[222] Loading up the parts – the footwells are filled. *Author*

[223] The hinges are worn, causing the doors to sag when open, so I have taped a note (partially seen) on them warning that they should be lifted before shutting. *Author*

[224] The car is packed full. *Author*

[225] A moment of reflection with my Ferrari. *Author*

[226-228] Moving day – my employees push 0046M out of the warehouse, onto the lift gate and into the transporter. *Author*

[229] The engine parts, transmission and fuel tank are brought out and loaded behind the Ferrari. *Author*

[230] The transporter drives away, leaving memories behind. *Author*

Chapter 11

2007-2017: EPILOGUE

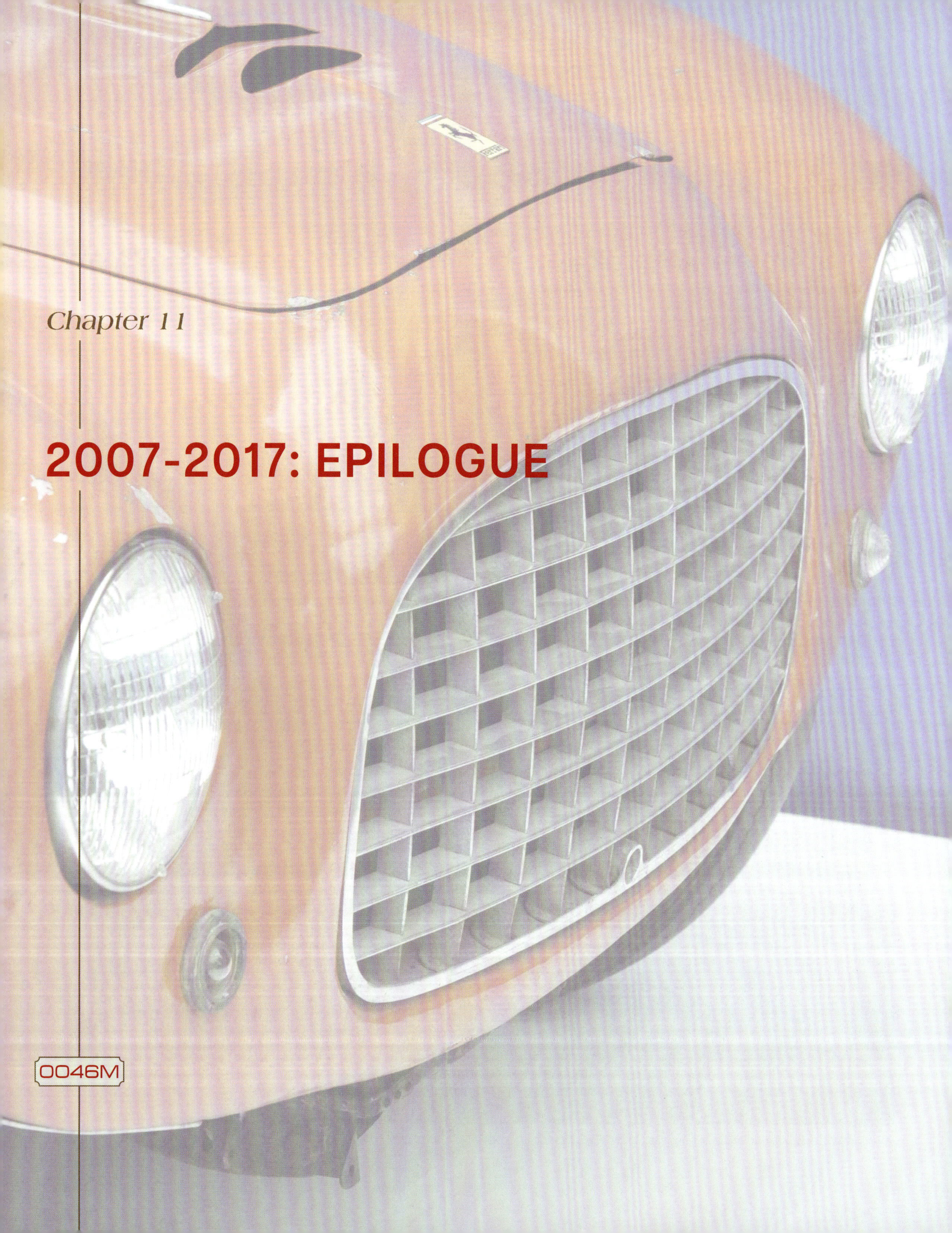

Even though I had sold it, I still regarded 0046M as my car and I wanted to maintain a connection with it, so I asked Micalizio to act as an intermediary and arrange for me to call Scott Gauthier so we could get acquainted. I was given his phone number and a time to call, and we spoke for a short time. True to what I was told, he was a serious collector specializing in Zagato and had an impressive number of significant cars. Gauthier was enthusiastic as I expressed my hope that the Ferrari would be restored and I offered to assist by sharing the information I had accumulated over the years for that purpose. He referred me to Rich Heinrich, one of the two mechanics who would be doing the work at the facility he and William Pope shared. This was a private museum and restoration shop where they kept their collections, named the Scottsdale Automotive Museum. Figure **[231]** shows Gauthier in his Fiat 750 Zagato.

I had a few more phone conversations with Gauthier and he was always cordial, but Rich Heinrich and I talked many times by phone and got to be friends of a sort. Rich was eager to get started on 0046M, but there were other projects he had to tend to that had priority. We talked about the details of the car, though, and went over issues I thought were important for him to know. For future reference, I sent him the templates to fabricate new motor mounts and the set of measurements and photographs I took at Bob Smith's shop to give him details on the drive shaft and right side axle tube that would be needed.

Rich could not resist a mock assembly of the engine to see if all was there and what it looked like. His photographs in **[232** and **233]** show it stacked together and the Ferrari sandwiched in with other rarities awaiting their turn. Since

[231] Scott Gauthier, the new owner, sits in his restored Fiat 750 Zagato. *Carl Schultz*

[232] Mock assembly of 0046M's engine at Scottsdale Automotive Museum, Arizona. *Rich Heinrich*

[233] The Ferrari and its engine among other cars in the restoration shop. *Rich Heinrich*

the engine needed considerable work, in early 2008 they sent it to Epifani Restorations in Berkeley, California, for assessment and repairs. In house, they fabricated new seat frames since the originals had fairly well disintegrated from rust due to southern humidity being retained in the fiber padding.

Meanwhile, Audrey Saffell and I were still in touch and it happened that she planned to visit an old friend in Birmingham – George Bernard, whose Formula Junior car she had raced in the early '60s. Naturally we wanted to meet, so on April 25, 2008, I made the hour drive over to Bernard's home and we had a wonderful time reminiscing about the Ferrari and looking at the photographs I had collected in an album. We posed together with the album and I photographed her with one of the 0046M models **[234** and **235]**.

Audrey lamented that she had missed the Ferrari by a year – had she come to Alabama earlier, she could have come to Tuscaloosa to see it. Since she was an avid traveler and her home in Southern California was not too far from Scottsdale, however, I eventually arranged for her to visit 0046M. Gauthier was amicable and Rich met her at the airport on March 1, 2010, then showed her around the museum and reunited her with 0046M. The last time she had seen the Ferrari was 50 years before, after Chuck had refurbished it, and it was now time worn and mostly dismantled, not to mention having the Chevy engine, so she was a little taken aback but still enjoyed seeing it and reviving old memories **[236** and **237]**.

From time to time as the years rolled by, I would call Rich and ask if anything was happening with the Ferrari and the answer was invariably the same: nothing. Just as it had done at my warehouse, it was sitting untouched.

At Epifani Restorations, though, they were slowly and carefully proceeding on the rebuild of the irreplaceable engine. Fortunately, copies of their invoices are available and they show how much work was needed and how much frustration was involved in sourcing parts, only to find that they were unavailable or incorrect when received. Even

[234] Audrey Saffell Sundin and I finally meeting in Birmingham, 2008. *Author*

[235] Audrey holds a model of our Ferrari. *Author*

[236] Audrey visits 0046M at the Scottsdale Automotive Museum, 2010. *Audrey Saffell Sundin*

[237] 0046M on display in Scottsdale. *Audrey Saffell Sundin*

special tools had to be made for some of the operations involved. As I said earlier, it was good that Bob Rabun and Mike Carpenter were not able to finish their rebuild, since it might well have led to further damage to the engine.

The engine block was the first to receive attention. To ensure against metal fatigue, all of the original studs were removed, both for the main bearing caps and the heads, together with all of the locating dowels, and the cylinder liners were taken out. New studs and dowels had to be custom made as were new cylinder liners. The block required some welding repairs and much machine work, and then the new liners were put in. New pistons, rings, rods, bearings, timing chain and gaskets were sourced and bought, along with many other small parts. The original crankshaft was deemed unsafe so a new crankshaft was made to order, but it turned out to be flawed, so a replacement was made. That too had issues, and it was not until a third new crankshaft was received that one was finally correct. The block was line bored and many trial assemblies were done and careful measurements made with gradual, minute corrections. All parts were magnafluxed to ensure there were no cracks, then polished or bead blasted.

Then came the cylinder heads. While Bob Rabun and Mike Carpenter were stymied by the difficulty of finding horizontal mousetrap-style valve springs for their rebuild nearly 50 years before, there are now conversions available to replace those on the early V12s with vertical coil springs, and when Epifani took the project they ordered that conversion. The valve springs were only one of the problems facing them with the heads, though. The 225-style heads 0046M has were described earlier with the more complicated roller bearing cam followers and the repairs Carpenter did to the damaged combustion chamber would never suffice to preserve such a rare and valuable engine. Plus, there were numerous cracks, gouges and ruined threads that had to be repaired. On top of all that, a head was warped, all evidence of the hard life of a historic racing engine.

Back in Scottsdale, Gauthier embarked on a new venture in 2015 that would both showcase his collection and make it profitable. He acquired a huge building at 602 N. Hayden Road in Scottsdale and opened G Collection, moving 0046M there to join his other cars from the Scottsdale Automotive Museum. As his website describes it:

> *G Collection is a rare and unique entertainment and dining venue that is sure to amaze and impress even the most discriminating clientele. This venue is an experience in entertaining that is unmatched anywhere in the world. The central focal point of this experience is a world-class classic automobile collection. A second underlining theme is an appreciation and respect for collectible antiques and rare, historically significant, antiquities.*[195]

So, 0046M was put on display in its dismantled state among pristine restored examples of automotive elegance

[195] *www.gcollectioncars.com/about.html*, accessed September 15, 2022.

[238-239] Scott Gauthier's G Collection venue in Scottsdale. *Scott Gauthier*

[ABOVE: 240] 0046M is the ugly duckling of Gauthier's collection. *Scott Gauthier*

[LEFT: 241] The display plaque has several exaggerations. *Scott Gauthier*

[238 and **239]**. It appears that some of the components were loosely stored inside the Ferrari and the temporary scrap steering wheel only made appearances worse **[240]**. I wonder what the chic clientele thought about the ragged little red car and if they realized that it was really the jewel of the collection. Perhaps that is why its plaque exaggerated what was already an exceptional history **[241]**.

As for why the Ferrari had remained unrestored since he bought it, Gauthier had recognized that 0046M was something special and took a different approach from what he had done with other cars. As the *Los Angeles Times* reported,

> *Gauthier considered restoring the car, perhaps to perform what is sometimes called a "sympathetic restoration," which would straighten bent body parts, remove rust and recreate original-looking upholstery without making it look museum-quality or movie-studio fresh. He decided against it, he said, out of respect for the car's lineage and its place in history.*
>
> *"This is probably one of the last great historic Ferraris that is not restored," Gauthier said. "I was worried that if I restored it, I might have been doing it a disservice."*[196]

His thinking mirrored a movement that had been underway for probably a decade as a reaction to all the over-restored cars being seen at Concours events. The competition was for perfection, to make cars better than their manufacturers had turned them out and that reconstruction belied their use through the years for the purpose they were built. In contrast, the preservation school of thought stresses leaving a car in its original condition, faded and worn, as a time capsule, and that is becoming a popular approach in collecting.

By early 2016, Gauthier decided to sell the Ferrari during the upcoming Monterey Car Week in August. The car was consigned to Gooding & Co. for their Pebble Beach Auction on August 20, so Gauthier's crew set about making 0046M presentable and the hope was to have the rebuilt V12 installed and the car running.

[196] *www.latimes.com/business/autos/la-fi-hy-pebble-beach-preservation-20160801-snap-story.html.*

Over at Epifani Restorations, work was still in progress on the engine. The head that had suffered the dropped valve required much attention, with material welded in and shaped, and the bore for the guide of that valve was damaged and had to be reworked. New rollers and bearings were found. The valve seats were removed and replaced with new ones. New valve guides were manufactured and installed. The pressure of a deadline began to show when new valves were ordered and that source could not produce them quickly enough. On the invoices there is a labor item for June 21: "Source new manufacturer of valves that can meet customer's schedule." Finally, all of the work was done, but my summary does not begin to do justice to all the effort and detail that went into the repairs over eight years. The final invoice was dated July 13 and the total of the invoices was $109,000.[197] The finished and crated engine block is shown in **[242]**.

However, Epifani Restorations did not assemble and fire up the engine; they only did the necessary repairs and ensured that everything was perfectly restored to the point where the engine could be assembled and started. That job fell to Gauthier's mechanic, who was faced with both a looming deadline and an unfamiliar engine to build, install in the car, and get running. The car was due at the auction site by August 17 for bidder viewing, and if the components were shipped from Berkeley around July 13, per invoice, then allowing transit time he might have had four weeks to do the work.

[197] In 2024 dollars: $142,460.

[242] The refurbished 0046M engine block as shipped by Epifani Restorations. *Copyright and courtesy of Gooding & Company*

Unfortunately, all did not go as planned. My understanding is that he could never get one bank of cylinders to operate. Out of time, the Ferrari had to be shipped and presented as non-running.

On the morning of July 20, I received an email from the university's Office of Advancement alerting me to a *Forbes* online article from the previous day about Gooding & Co.'s offerings for the Monterey auction, one of which was 0046M. Immediately I went to work booking a flight, a rental car and accommodations somewhere near Monterey, which was not easy at such a late date since Car Week draws numbers in the hundreds of thousands, but by the end of the day I had all three. I also called Gooding & Co. and spoke to Justin Gosaynie, a sales specialist, explaining that I was the fellow they were mentioning in their press releases who had bought 0046M while in college and kept it so long. I went on to say that I was coming out to watch the auction and would like to visit my car again; my unsaid mission was to hopefully meet the new owner. Gosaynie was graciously accommodating and said I would be most welcome; all I needed to do was ask for him at the registration desk. I had not talked to Rich Heinrich in some time, so I called him and he filled me in on some of the developments related above.

The next day, the 21st, I had an email from my friend Randy Cook in Tallahassee who had heard about my car being auctioned and wanted to notify me. Randy had been a student at the University of Alabama the year I had the Ferrari on campus and had seen and admired it although we never met back then. It was 50 years later when Randy was working on his book *Bowtie Ferraris*, cataloging the Ferraris that had received Chevy engine conversions, that he got in touch with me and we realized that unknowingly our paths had crossed many times since and that we had many acquaintances in common. After that we became good friends. Randy said he was going to Monterey, too, to promote his new book, *Blue Oval Ferraris* (with Ford engines). Coincidentally, we were both staying in the town of Gilroy, so we got together for dinner on the night of August 18 when we arrived.

The following morning, I drove in to Monterey and found the Gooding site. After the staff located Gosaynie, he greeted me and ushered me in. Right away I could see 0046M, sitting on a raised platform in a place of honor as one of the featured cars **[243 and 244]**. We walked past rows of fabulous cars, but my eyes were on my baby – last seen nine years before. It had been waxed and polished, over paint that had long been chipped and cracked in places, but the old familiar lines were there, still with the flaws from my amateur bodywork. And there was my University of Alabama parking decal, still on the windshield. Gosaynie opened the driver's door for me and I had to shake my head – the seats had been reupholstered in black leather. Apparently, it had been a case of artistic license to match the dash that had been accurately recovered in the same. Someone's taste trumped originality, I suppose, but then nobody bothered to ask what the original was. I noticed as well that they had used the incorrect extension shifter top on the transmission and had fabricated a cover to fit that, since the concern was only to display a complete car for sale **[245]**.

I also saw that they had, in haste I suppose, crudely glued the old red naugahyde headliner directly to the underside of the top and that left edges showing. They had managed to reuse some of the old matting I had saved that covered

[ABOVE & OPPOSITE TOP: 243-244] Gooding & Company auction, Pebble Beach 2016. The Ferrari is singled out on a platform above the lesser consignments. *Richard Michael Owen*

the interior contours of the rear wheel wells and the spare wheel deck, then found simple black carpeting for the floor, firewall and transmission cover **[246]**. The spare wheel itself was oddly missing.

On the dashboard, they had installed the original-style Magneti Marelli key switch and the lighted starter pushbutton I had acquired, but the old key switch was still wired and hanging down under the dash, so I doubted if those were operable. Since the central switch panel under the dash

[245] The dashboard has been recovered in black leather but so have the seats. The extension shift top has been used on the transmission. *Richard Michael Owen*

[246] The tattered original Zagato matting has been returned to the spare tire area, but no spare tire. Leopold's red headliner has been roughly attached. *Richard Michael Owen*

had not been replaced either, my impression was that the electrical system required much work.

On the outside, it was the same as when it left my shop, except that they had added the Zagato name below the filler cap to emulate the way it was when Bosisio had the car, and there was now an old California tag. The rare cap for the hand crank hole in the bottom of the grille was missing, but I attributed its absence to a concern that as a loose part it could be stolen.

Of course I was eager to see the engine. When I asked Gosaynie if I could open the hood, he said he would do it. I knew he meant well and was used to protecting cars of astronomical value from the hoi polloi who pass through on viewing days, but I felt sad about that, not indignant, only sad that times had changed and it was no longer just my old car anymore. But then I saw that magnificent V12 filling the engine bay at last and I was smiling.

Anyone who appreciates the beauty of machinery has to love Ferrari engines. There is a marvelous complexity to them. All the eye candy of chrome and polished and matte aluminum in sensuous castings, abundant wires, hoses and assorted gadgetry – all conspire to delight the senses and exude power while sitting there **[247-249]**. On closer examination, though, I saw that the carburetor linkage to the gas pedal was missing and there was no connection to the radiator. In essence, it was a static display, but I already understood it could not run. I noticed, too, that one of the riveted Weber name plates on the air cleaner tops had been removed – liberated, I presumed. The brown lifting knobs on each end of the valve covers were missing as well, but I thought that might also be for fear of theft. I saw that the generator was not the original but a similar replacement with a black band around the brushes instead of a chromed one. Conversely, the new regulator box above the generator was now silver instead of black. On the plus side, the old regulator on the firewall that went with the Chevy generator was gone, and I was happy to see that they – Epifani probably – had found a pair of original-type mechanical fuel pumps. At some point, the two FISPA mechanical pumps had been removed and replaced by a single Bendix electric pump and their mounting holes covered by plates.

[247-249] The V12 has been installed but is not running.
247: Richard Michael Owen, 248/249: Author

Sadly, I was to learn later that the parts I noticed missing were lost. I was particularly perturbed that the rare cap for the hand crank hole I had guarded zealously through the years had disappeared.

I spent several hours hanging around the car and talking to people who stopped and took an interest in it, especially my student parking decal, which was a real attention-getter. I approached one fellow with some serious camera equipment who was taking photographs of the car and asked if he worked for a magazine. He did, and said he was David Tomaro from *Sports Car Market*. I explained who I was and that I had only a small camera, so could I possibly get copies of his photographs later? He said he would be glad to send copies to me and asked me to pose with the Ferrari. Later on, his magazine printed that photograph in their November 2016 issue on page 62 **[250]**.

The auction would not begin until the evening of the 20th, so that morning I joined Randy at the Concorso Italiano show where he had a table set up among the vintage Ferraris to sell his books. We had a great time meeting people, talking about Ferraris and even selling a few books. Toward the end of the day we broke camp and it was time for me to say goodbye to Randy and head over to the Gooding site. I never saw him again; six months later Randy died from a fast-moving cancer.

At Gooding's giant auction tent, I found a chair near the aisle in the rear spectator seating area and prepared for a long wait. 0046M was lot number 68 so it would be quite some

[250] I stand beside 0046M for a photograph by David Tomaro, published in *Sports Car Market* magazine. *David Tomaro*

time before its turn on the block. Meanwhile, a father and his adult son had taken seats beside me. They were visitors who had come for the spectacle and one asked me, in jest, if I was there to buy a car, to which I replied, no, I was there to see one sold. They asked if I was selling the car and that led to my giving them a sketch of what had transpired. They were fascinated and I think it gave them a sense of participation rather than being just bystanders.

As the evening wore on, we watched as car after car was bid and sold, some going for staggering prices, others almost affordable, but the auction was entertaining, especially with the dry humor of the auctioneer, Charles Ross. Even with its lighter moments, there was an atmosphere of gravity and decorum, quite a contrast to the circus nature of the muscle car auctions I had seen on television. This was a spectacle, but a refined one.

[TOP: 251] Auction time and 0046M has been pushed on stage for the opening bid. *Angus MacKenzie*

[RIGHT: 252] Sold for $4,950,000, plus commission. *Author*

Finally, Lot 68 was pushed onstage and into the spotlight. I stood and moved out in the aisle so I could take photographs and be able to see who was bidding, thinking I might be able to go introduce myself to the winner at the conclusion of the evening. After a glowing synopsis of 0046M's pedigree by David Gooding, Ross opened the bidding at $2,000,000, which was quickly met **[251]**. It rose gradually, one or two hundred thousand at a time, but I was too far away to see who the bidders were. At first, there were a number of bidders but by $3,000,000 there were three and soon one dropped out. At $4,800,000, another bidder joined in. The bidding slowed considerably from that point with each increment being carefully considered. Finally, at $4,950,000, two of the remaining three bidders dropped out and the car was hammered sold **[252]**. With Gooding's 10% commission, the total came to $5,445,000.[198]

My father and son seatmates were agog at the money changing hands for my old car. The son asked me how I felt about it. I told him I had no regrets or envy because that was the way things worked – just as when you sell a house and years later the buyer sells it again for more money.

The winning bidder was somewhere diagonally opposite me and near the front, so I had no idea who to look for. It was getting late by then and I had a long drive in the dark back to my hotel in Gilroy. Besides, I felt sure Gooding would put me in touch with the new owner of my Ferrari, so I left. Early the next morning I was on my flight back home.

Giving the Gooding crew some time to regroup, I emailed Gosaynie on August 30, asking about contacting the buyer to let him know I had the car's history. He said he would forward my request and that I would probably get a response. Sure enough, two days later I had an email:

> *My name is Rainer Wolf and I am responsible for Mr. Pohl's cars, who is the new owner of the beautiful*

[198] In 2024 dollars: $7,116,600.

> *0046M. Mr. Pohl was very pleased to hear that you took a lot of effort throughout your ownership to collect history. If this would be ok for you, I would like to call you to discuss further details.*

On the one hand, I was elated he had reached out to me, on the other hand I was dismayed when a quick search for "Pohl Ferrari" confirmed my suspicion. Andreas Pohl was the new owner and the car would be in Germany. It could be that I would never see it again. Pohl was in good position to take care of it, though, since according to *Forbes* he was the co-owner with his brother of Deutsche Vermögensberatung (DVAG), the leading German company offering consumer insurance, annuities and mutual funds. I learned that he had sponsored seven-time Formula One World Champion Michael Schumacher throughout his career with Ferrari and after his tragic accident has continued to assist him and his family as well as sponsoring his son in racing. He is an avid car enthusiast with a huge, eclectic collection and often enters concours and vintage driving events. Among them, Pohl and his daughter Nathalie participated in the 2016 Mille Miglia reenactment with his 275 Sport/340 America Ferrari, 0030MT **[253]**.

I replied to Rainer and a few days later received an email from his secretary, Brigitte Astor, who wanted to coordinate a phone call between Rainer and me. It turned out that Rainer travels constantly on Pohl's behalf buying, displaying and exercising the cars in his collection. Eventually, we connected and had a long and rewarding conversation. When I asked what their plans were for the Ferrari, Rainer explained that the philosophy of their collection was preservation rather than immaculate restoration. The plan for 0046M, then, was to eventually get it mechanically functional but otherwise

[253] New owner Andreas Pohl and his daughter Nathalie in the 2016 Mille Miglia recreation. *Massimo Matteucci*

leave it alone. I told Rainer about my work researching the history of 0046M and he was impressed that I knew so much about the Ferrari. As we closed the call, I had the feeling that my car and I were among friends.

The following March, in 2017, I had an email from Ms. Astor, asking if I would be going to the Amelia Island Concours. Her reason for asking was that Rainer would be there and would like to meet me. When I told her that I was not going, she said in that case Rainer would fly up to see me. That, I thought, was a gesture in keeping with people who are genuinely involved with their cars.

I met Rainer at the Birmingham airport after his flight had been long delayed by weather, and drove him to the historic Redmont Hotel, where I had suggested Ms. Astor make a reservation. Once he had settled in, we met in the restaurant **[254]** and began a delightful evening of getting to know each other, talking about Andreas Pohl's amazing collection that he managed (260 cars, at that time) and in particular 0046M. I gave him a sketch of how I bought the Ferrari and what I had done, showed him the photograph album I had amassed, then for the climax I presented my three-inch thick binder of documents and correspondence representing decades of research. Pouring through everything, Rainer was fascinated by what he saw. I explained to him that I had done all that with the intention of writing a book on the history of 0046M someday. Generously, Rainer offered to sponsor its publication, and that is how this book came to be.

[254] Rainer Wolf, curator of Pohl's collection, meeting me in Birmingham, Alabama. *Author*

Rainer flew out the next morning, tending to the business of Pohl's ever-expanding collection, but he left me with a photograph of 0046M in its new sanctuary **[255]**, a huge underground garage at Pohl's home where he keeps his best cars. I saw it in repose with its stablemate, 0052M, another 166 Mille Miglia but in its original *barchetta* form, also preserved with the patina of time and use. My Ferrari is in good company there, as shown in a perspective of the class that surrounds it **[256]**.

In the book *Traumgaragen 2.0 Deutschland* (German Dream Garages 2.0), the space is described as being "about 55 meters long and 20 meters wide and there are about 30 vehicles in it. Together with an architect friend, the collector came up with the underground car park project. A spacious room above it and at ground level offers space

[255] 0046M beside 0052M in Pohl's underground garage. *Rainer Wolf*

for automobilia, literature, a small bar and meetings with friends."[199] The window at the far end is actually a glass panel set into the side of his swimming pool. Judging from Pohl's smile **[257]**, it is obvious that he is pleased to have become one of us.

For me, it is reassuring to think that in Pohl's hands, at least, the amateur bodywork of my youth will remain and the paint my brother applied will be a kind of memorial to him now that he is gone. 0046M will outlast us all and there will be many more chapters to add to its story. Someday the V12 will be running again, the shifter will be as it was, and hopefully the black seats will be gone. Maybe I will be around to see that, and if so, maybe I will get a chance to drive it one more time.

I'd like that.

[199] Fritz Schmidt, Jr., p. 123.

[LEFT: 256] Pohl's home garage houses only a fraction of his collection. The window at the far end is a glass panel set in the side of his swimming pool. *Traumgaragen 2.0 Deutschland*

[BELOW: 257] Andreas Pohl, happy with his historic Ferrari. *Rainer Wolf*

IN MEMORIAM

Enzo Ferrari (1898-1988)

Giuseppe "Nuccio" Bertone (1914-1997)

Emilio Maria Giletti (1929-2020)

Luigi Attilio Bosisio (1924-1998)

Bruno Martignoni (1913-1986)

Luigi Chinetti (1901-1994)

David Francis Leopold (1924-2005)

John Reynolds Sharpe (1926-2006)

Delavan S. Lee (1914-1978)

Charles "Chuck" W. Saffell (1918-2006)

Audrey Saffell Sundin (1935-2021)

Maurice Kevion Bivens (1934-1991)

Charles G. Deerman (1935-2009)

BIBLIOGRAPHY

Acerbi, Alessandro and Massimo Acerbi	*L'Aosta: Gran San Bernardo: 1920-1957*	Aosta, Italia: La Vallée, 1984
Alfano, Nello	*La Cagliari-Sassari-Cagliari*	Cagliari, Sardinia: Serra Editore, 2010
Amati, Luigi	*50 anni di automobilismo sportivo in Puglia*	Fasano, Italia: Grafischena, 1976
Anderloni, Carlo Felice Bianchi	"Barchetta... a Myth."	Registro Internationale Touring Superleggera, April 1998
__________ and Angelo Tito Anselmi	*Touring Superleggera: Giant Among Classic Italian Coachbuilders*	Roma: Autocritica, 1983
__________ and Valerio Moretti	*Ferrari 166 MM Barchetta*	Milano: Automobilia, 1991
Anselmi, Angelo Tito and Lorenzo Boscarelli	*Ferrari Tipo 166: The Original Sports Ferrari*	Somerset, England: Haynes Publishing Group, 1986
	Associazione Italiana per la Storia dell'Automobile	*Monografia 46, 55*
	Auto Italiana	1950-1954
Bäumer, Walter	*Maserati A6G 2000 Zagato*	Deerfield, IL: Dalton Watson Fine Books, 2013
	Birmingham News	Birmingham, AL, 1957, 1962
Box, Rob de la Rive and Richard Crump	*The Automotive Art of Bertone*	Somerset, England: Haynes Publishing Group, 1984
	Buffalo Courier-Express	Buffalo, NY, 1956-1957
Cancellieri, Gianni and Cesare De Agostini	*Polvere e Gloria: la Coppa d'Oro delle Dolomiti (1947-1956)*	Milano: Giorgio Nada Editore, 2000
Candini, Dante and Nunzia Manicardi	*Stanguellini: il "mago" dei motori*	Modena, Italia: L'Olmo e la Vite, 1999
Castellarin, Danilo	*Correndo nel Tempo: Storia del motorismo veronese nel Novecento*	Verona, Italia : Cierre Edizioni, 2017
Cavallino	John Barnes, Jr., Publisher	Palm Beach, FL
Colombo, Gioachino	*Origins of the Ferrari Legend*	Somerset, England: Haynes Publishing Group, 1985
Cook, Randy	*Bowtie Ferraris: Chevy-Engined Ferraris from the 1950s and 60s*	Tallahassee, FL: RAC Motorsports Publishing, 2014
Costantino, Augusto, ed	*Ferrari Catalogue Raisonne 1946-1986*	Milano: Automobilia, 1987
Curami, Andrea, ed	*Mille Miglia Race: The Postwar Years*	Milano: Giorgio Nada Editore, 1998
Dal Monte, Luca	*Enzo Ferrari: Power, Politics and the Making of an Automotive Empire*	Phoenix, AZ: David Bull Publishing, 2016
Davis, David E., Jr	"Remembering Del Lee: Allards, Ariels and Atilla the Hun."	*Car and Driver*, May 1978
Dregni, Michael	*Inside Ferrari*	Osceola, WI: Motorbooks International, 1990
	Ferrari Market Letter	Cathy Roush; Suwanee, GA

	Ferrarissima	Milano
Fondi, Pino	*Il mitico Giro di Sicilia*	Milano: Giorgio Nada Editore, 1996
Froissart, Lionel	*Ferrari Pininfarina*	New York, NY: Universe Publishing, 1997
Gasich, W.E	*Forty Years of Ferrari V-12 Engines*	Warrendale, PA: Society of Automotive Engineers, Inc., 1990
Giuseppe, Prisco	*Coppa della Perugina: giro automobilistico dell'Umbria*	Torino: Edizioni ASI Service S.r.l., 2014
Grayson, Stan, ed	*Ferrari the Man the Machines*	Princeton, NJ: Princeton Publishing, Inc., 1975
Greggio, Luciano	*Bertone*	Milano: Giorgio Nada Editore, 1992
Hamilton, Duncan	*Touch Wood*	London: Speed and Sports Publications, Ltd., 1971
Lurani, Giovanni	*Mille Miglia 1927-1957*	*Switzerland: Automobile Year, 1981*
Marchianò, Michele	*Ferrari by Zagato*	Milano: Giorgio Nada Editore, 1988
__________	*Le Zagato: Fiat 8VZ-Alfa 1900 SSZ*	Milano: Libreria dell'Automobile, 1987
__________	*Zagato: Seventy Years in the Fast Lane*	Milano: Giorgio Nada Editore, 1989
Marzoli, Robi	*Martignoni l'africano*	Varese, Italia: Alfatherm, 1991
Marzotto, Giannino	*La Ferrari alla Mille Miglia*	Roma: Autocritica, 1987
__________ *and Sergio Cassano*	*Red Arrows: Ferraris at the Mille Miglia*	Milano: Giorgio Nada Editore, 2001
Massini, Marcel	"Ferrari 166 MM Barchetta Touring 1950, Chassis #0046M."	2016
Merritt, Richard F	*Ferrari: Brochures and Sales Literature – A Source Book 1946-1967*	Scarsdale, NY: John W. Barnes, Jr., Publishing, Inc., 1976
Millanta, Corrado, Luigi Orsini, and Franco Zagari	*Ferrari Automobili 1947-1953*	Firenzie: Editoriale Olimpia S.p.A., 1985
Moretti, Dario	*Gilco: Le macchine di Gilberto Colombo*	Milano: Trafiltubi Srl, 1993
O'Neil, Terry	*Northeast American Sports Car Races 1950-1959*	Dorchester, England: Veloce Publishing Limited, 2010
O'Shea, Paul	*Guide to Competition Driving*	New York: Sports Car Press, Ltd., 1957
Oostheok, Willem	*Sports Car Racing in the South: Texas to Florida 1957-1958*	Deerfield, IL: Dalton Watson Fine Books, 2011
Orsini, Luigi and Franco Zagari	*OSCA: la rivincita dei Maserati*	Milano: Giorgio Nada Editore, 1989
Perna, Luigi	"La Storia Emilio Giletti."	*Sportweek*, September 2019
	Primavera Bresciana XX e XXI Coppa F. Mazzotti Mille Miglia	Brescia: Prodotto Nazionale e Commercio Estero, 1954
Quadraroli, Benito	*Circuito di Senigallia*	Senigallia, Italia: Minardi Petroli, 1989

Reggiani, Francesco and Keith Blumel	*Ferrari Engines: Enthusiasts' Manual*	Somerset, UK: Haynes Publishing Group, 2018
Scaletti, Francesco	*Ferrari I Quattro Cilindri*	Roma: Mito Rosso, 1998
	SCCA Membership Directory	1956, 1958
Schmidt, Jr., Fritz	*Traumgaragen 2.0 Deutschland*	Rüsselsheim, Deutschland: Fritzclassics, 2012
Tanner, Hans	*Ferrari Owner's Handbook*	Los Angeles: Floyd Clymer Publications, 1960
__________ and Doug Nye	*Ferrari, 6th Edition*	Somerset, UK: Haynes Publishing Group, 1986
Valeriani, Frederico	*Coppa Acerbo: Circuito Internazionale Degli Abruzzi*	Edizioni Res Imago, 2008
www.archivioluce.com		
www.barchetta.cc		
www.cameva.it		
www.commons.wikimedia.org/wiki/category:Public_domain		
www.exchangerate.com/currencyinformation/italian-lira		
www.ferrarichat.com		
www.gcollectioncars.com/about.html		
www.lanciamarino.it/sito/index.php/fr/contexte-historique/circuits		
www.latimes.com/business/autos/la-fi-hy-pebble-beach-preservation-20160801-snap-story.html		
www.motorsportmagazine.com/database/drivers/giuseppe-farina		
www.racingsportscars.com		
www.sharpenet.com/give-take/commitment-education-drives-gifts/		
www.targapedia.com		
www.usinflationcalculator.com		
www.youtube.com/watch?v=6Jcg04CCw5M		
www.youtube.com/watch?v=CnlSdrTcnVM		
www.youtube.com/watch?v=gVxmGXUpYU0		
www.youtube.com/watch?v=RKJ-kO80eHw		
www.vintageracecar.com/watkins-glen-1956/2/		
www.wikipedia.org/wiki/Italian_lira		
Zagato, Elio	*Elio Zagato: storie di corse e non solo*	Milano: Giorgio Nada Editore, 2002

Appendix 1: Early 166 Mille Miglia Ferrari Sales Brochure, page 1

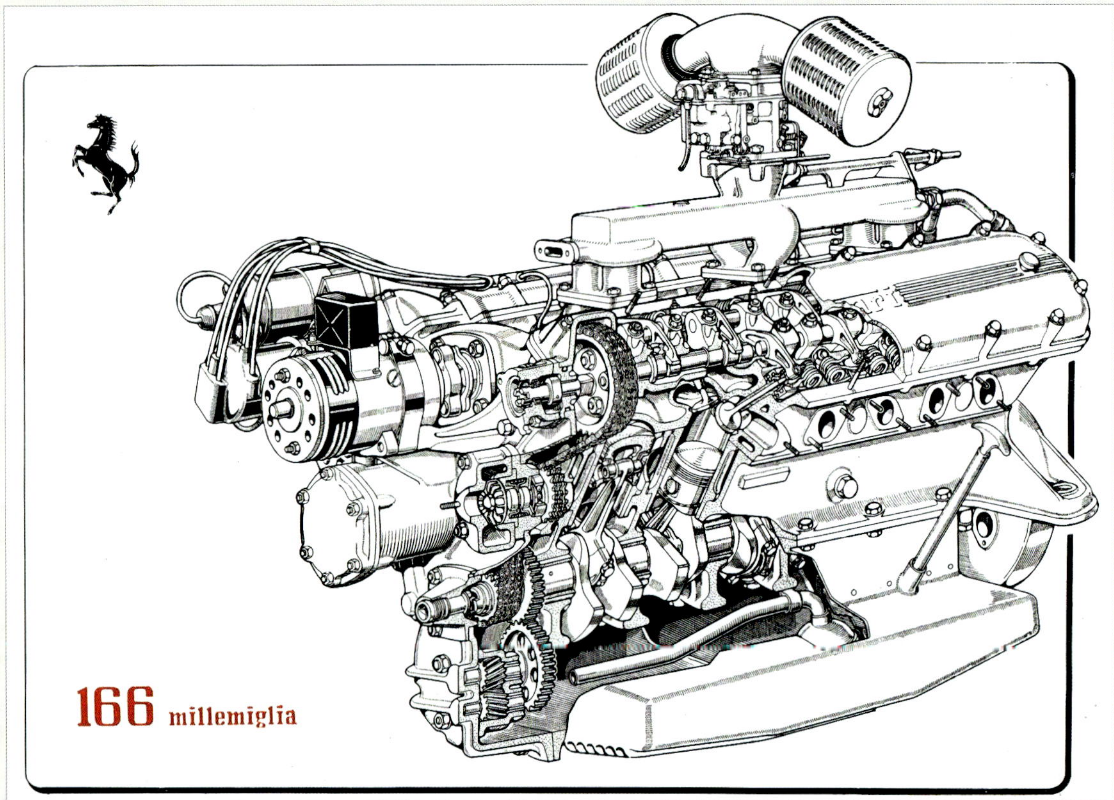

Appendix 1: Early 166 Mille Miglia Ferrari Sales Brochure, page 2

GENERALITÀ

MOTORE

Numero dei cilindri disposti a V con apertura a 60°	12
Alesaggio e corsa	60 x 58,8
Cilindrata unitaria cmc.	166,25
Cilindrata totale cmc.	1995
Rapporto di compressione	8
Potenza massima Cv.	130
Regime corrispondente	7000
Blocco cilindri e basamento in un sol pezzo in silumin con canne riportate in ghisa speciale	
Testa in lega leggera con camere di scoppio emisferiche	
Albero motore contrappesato appoggiante su 7 supporti	
Bielle affiancate su bronzine	
Distribuzione a valvole in testa a V leggermente sfasate e comandate da un albero a camme per ogni testa, azionato con catena silenziosa, regolata da tenditore automatico	
Bilanciere con registro di giuoco in corrispondenza della valvola	
Lubrificazione a pressione con pompa ad ingranaggi	
Radiatore per olio e coppa in silumin	1
Accensione a batteria e distributori con anticipo automatico	2
Alimentazione con pompe autoregolatrici a membrana	2
Carburazione con carburatore a doppio corpo invertito **Weber** 36 DCF	1
Piazzamento elastico su 4 tamponi di gomma	
Frizione a disco unico con mozzo elastico	

AUTOTELAIO

Cambio in blocco col motore, a 5 velocità tutte comandate con leva, 2ª 3ª 5ª silenziose, 4ª in presa diretta	5
Ponte posteriore del tipo rigido leggerissimo	
Telaio monoblocco in tubi d'acciaio a sezione ellittica	
Sospensione anteriore a ruote indipendenti con balestra unica trasversale ed ammortizzatori Houdaille a regolazione termostatica	
Sospensione posteriore a balestre lunghe semiellittiche, con triangolo di reazione ed ammortizzatori Houdaille	
Sterzo con parallelogramma articolato ed indipendente dalle oscillazioni delle ruote. Guida a destra con vite senza fine e ruota elicoidale	
Freni idraulici sulle 4 ruote con comando a pedale, meccanici sulle ruote posteriori con comando a mano	
Passo - Carreggiata	2,25x 1,27a / 1,25p
Prestazioni velocità massima della autovettura a pieno carico	km/ora 210
Peso a vuoto dell'autovettura carrozzata spyder	kg. 680
Consumo carburante *Esso* extra per 100 Km.	litri 18
Capacità serbatoio	litri 100
Ruote a raggi in lega leggera con mozzo Rudge	4,00 x 15
Pneumatici - PIRELLI	5,50 x 15

MODENA **Ferrari** ITALIA

Ferrari

II° G. P. INTERNAZIONALE DI TORINO - 1. Sommer

G. P. DI STOCCOLMA - SVEZIA - 1. Biondetti

12 ORE DI PARIGI - MONTHLERY - 1. Chinetti

G. P. FORMULA 2 - FIRENZE - 1. Sommer

AUTODROMO DI MONTHLERY

Chinetti - Record mondiale dell'ora alla media oraria di Km. 202,800
- Record internaz. dei 200 Km. » » » » » 203,378
- Record internaz. delle 100 miglia » » » » » 203,435

XVª VERMICINO - ROCCA DI PAPA - 1. Bracco

III° GRAN PREMIO DI ROSARIO - 1. Farina

VIII° E IX° GIRO DELLA SICILIA - 1. Biondetti

XVª E XVIª MILLEMIGLIA - 1. Biondetti

COMO - LIETO COLLE - 1. Bracco

GRAN PREMIO DI BRUXELLES - 1. Villoresi

VARESE - BRINZIO - 1. Bracco

GRAN PREMIO DEL LUSSEMBURGO - 1. Villoresi

COPPA INTER-EUROPA - 1. Sterzi - 2. Bianchetti - 3. Cornacchia

G. P. DI ROMA - 1. Villoresi - 2. Taruffi - 3. Cortese - 4. Righetti

II° GRAN PREMIO DI BARI - 1. Landi

III° G. P. DI BARI - 1. Ascari - 2. Cortese - 3. Bonetto - 4. Landi

GRAN PREMIO DI NAPOLI - 1. Vallone - 2. Bonetto

GRAN PREMIO DELL'AUTODROMO - MONZA
1. Fangio - 2. Bonetto - 3. Ascari - 4. Landi - 5. Cortese

24 ORE DI LE MANS - 1. Chinetti

II° GIRO DELL'UMBRIA - 1. Vallone

GRAN PREMIO DELLA SVIZZERA - 1. Ascari - 2. Villoresi

IX° CIRCUITO DEL GARDA - 1. Villoresi - 2. Tadini

GRAN PREMIO DELLE 24 ORE DI SPA - 1. Chinetti

IIª COPPA PICC. CILINDRATE - REIMS - 1. Sommer

IIIª COPPA PICC. CILINDR. - REIMS - 1. Ascari - 2. Tadini

IIIª COPPA DELLE DOLOMITI - 1. Vallone - 2. Cornacchia

GRAN PREMIO D'OLANDA - ZANDVOORT - 1. Villoresi

GRAN PREMIO DI SILVERSTONE - 1. Ascari - 3. Villoresi

CIRCUITO DI SENIGALLIA - 1. Cornacchia - 3. Stagnoli

CORSA IN SALITA - FASANO - SELVA - 1. Vallone

AOSTA - GRAN SAN BERNARDO - 1. Bracco - 2. Vallone

COPPA S. GIUSTO - 1. Cornacchia

G. P. D'EUROPA - MONZA - 1. Ascari - 5. Sommer

PONTEDECIMO - GIOVI - 1. Bracco - 3. Vallone

G. P. AUTOMOBILISTICO DELLA CECOSLOVACCHIA
1. Whithead - 3. Cortese

CORSA IN SALITA - BIELLA - OROPA - 1. Bracco

Telegrammi: Scuderia Ferrari Modena - Posta 232 - Telefono 4081 - Officine Telefono 94 - Maranello

Appendix 2: Assembly Sheets for 0046M, page 1

(Performance data redacted by Ferrari.)

Ferrari

Telaio tipo MM.50 matricola 0046 M

Motore tipo 166 M matricola 0046

Cambio tipo 166 M matricola 8

Ponte tipo 166 M matricola 0018

Collaudato il .... Consegnato il ....

Cliente ....

Appendix 2: Assembly Sheets for 0046M, page 2

Autotelaio tipo 166 M Motore tipo 166 M Matricola N.° 0046

FOGLIO MONTAGGIO MOTORE

Basamento silumin Coppa olio silumin Filtri 2

Albero motore nitrurato per Vandervell

Pistone Mondial Rapp. 8.5 Peso 181

Anello tenuta 125 Trione Raschiaolio 125 Trione

Bielle 166 Peso 441 Pompa acqua 166

Teste cilindri 166 normali Coperchi silumin

Guarnizioni teste anelli amianto c rame e klingerite 1000

Valvola asp. tipo 125 Valvola scar. tipo 125

Molla richiamo valvola Ø 3.2 Alberi distribuzione 166

Scatola distribuzione 166 silumin

Pompa mand. olio ingr. elicoidali Pompa di recupero no

Pompa benzina FISPA 2 Filtro benzina con le pompe Pres. aria 2 Weber D2

Carburatore tipo Weber 36 DCF/1 N. 1

Accensione con spinterogeni tipo ST 66 DAS Fase: D = $\frac{AA}{CS}$ 25/53 16/63 = $\frac{AA}{CS}$ 25/51 17/63

Frizione tipo "Confort" Carico 31

Giuochi: Albero-motore 0.075 Punteria 0.20-0.20

PRIMO MONTAGGIO

Data inizio montaggio Data fine montaggio 16/3/50 Montatori Neri G. Del Vacchi

OSSERVAZIONI

Spingidisco con piattelli saldati tipo 125.

Data 16/3/50 Il Capo Reparto F.to Franchini

OSSERVAZIONI DEL PRIMO RODAGGIO

Il Capo Reparto

SECONDO MONTAGGIO

Data inizio lavoro Data fine lavoro Montatori

OSSERVAZIONI

Data Il Capo Reparto

Appendix 2: Assembly Sheets for 0046M, page 3

RELAZIONE SALA PROVA

Motore tipo 166 N.° 0046M — Data 18/3/50

PRIMO RODAGGIO

Pressione barometrica .... Temperatura 16°

Umidità r. non rilevata Coefficiente di corr. ....

Carburante 80 ottani — Carburatore 1x36 DCF — Lubrificante Racer 50

Candele Marelli 300 A — Taratura: Diff. 25 — Getto 135 — Minimo 65 — Getto pompa 70

Freno aria 200 — Tappo-Spillo 225

Consumo olio = Gr./HP ora .... Consumo carburante = Gr./HP ora ....

Note di funzionamento Sul motore è montato [illegible] pozzetto speciale (vedi verbali S.P.S. pag. 25).

Motoristi: Marianelli, Vandelli

Entrato in S.P. il .... ore ....

Uscito di S.P. il .... ore ....

Appendix 2: Assembly Sheets for 0046M, page 4

Autotelaio tipo 166 M Matricola N. 0046 M

FOGLIO MONTAGGIO CAMBIO

Cambio tipo 166 M/50 Matricola 8

Scatola e coperchi silumin 166

Marcie 5

Sincronizzatore no

Denti ingr. prim. II ... III ... V ... rinv. II ... III ... V corta

Denti albero rinvio N.

Rinvio ad angolo per contachilometri 9x42

Pompa lubrificazione si

NOTE Completo.

Data 29/3/50 Il Capo Reparto F.to Franchini

Autotelaio tipo 166 M Matricola N 0046 M

FOGLIO MONTAGGIO PONTE

Ponte tipo 166 M/50 Matricola 0018

Coppia conica 8x40 Modulo corona 5.25

Planetari autobloccante

Satelliti autobloccante

Autobloccante si

Semiassi 166

Bracci laterali 166 normali

Flange di attacco 166

Scatola del ponte silumin

NOTE

Completo di tamburi freno tipo 400.

Data 21/4/50 Il Capo Reparto F.to Franchini

Appendix 2: Assembly Sheets for 0046M, page 5

Autotelaio tipo MM.50 Motore matricola N. 0046.M Telaio matricola N. 0046/M

FOGLIO MONTAGGIO AUTOTELAIO

Trasmissione MM.50

Freni anter. 400 - poster. tamburi 400 - suole 60 mm.

Mozzi e ruote mozzi Rudge - ruote 72 raggi 15 x 4.00

Pneumatici n° 5 - Pirelli

Ammortizzatori Houdaille

Sterzo 166 con giunti Fabbri

Radiatore acqua e olio 166 FIN MM.

Serbatoio carburante tipo MM. lampiera piombata capacità lt.120

Indicatore di livello 12 V.

Marmitta di scarico n° 2 MM.

Batteria Magneti Marelli 12 V. tipo 6VX9

NOTE Ponte rap. 0018 rap. 9x42 M/B
Cambio rap.166 M/B - n°8 - 9x42
Completo impianto elettrico - n° 2 avvisatori - freno a mano - acceleratore - piantone guida mm. 1405 - n° 1 motorino avviamento.

Note 19-4-52

Sono state montate N°2 Teste 225/16658
Albero distribuzione 410/16662-63
Bilancieri e molle 410-16641-42
Valvole asp. 400-16135
" scarico 225-16676
Molle valvole 225-16670
Molle bilancieri 225-16533
Albero motore con prima [illegible]
[illegible] - bielle

Vedi Commissione 474-C

Data 4/7/50 Il Capo Reparto F.to Levizzani

Appendix 2: Assembly Sheets for 0046M, page 6

Autotelaio tipo Motore matricola N. Telaio matricola N.

Data	REVISIONI E AGGIORNAMENTI
23/3/54	Trasformato motore a tre carburatori. Montato rapporto al ponte 8/40

Appendix 3: Bertone Purchase and Giletti Registration Turin, page 1

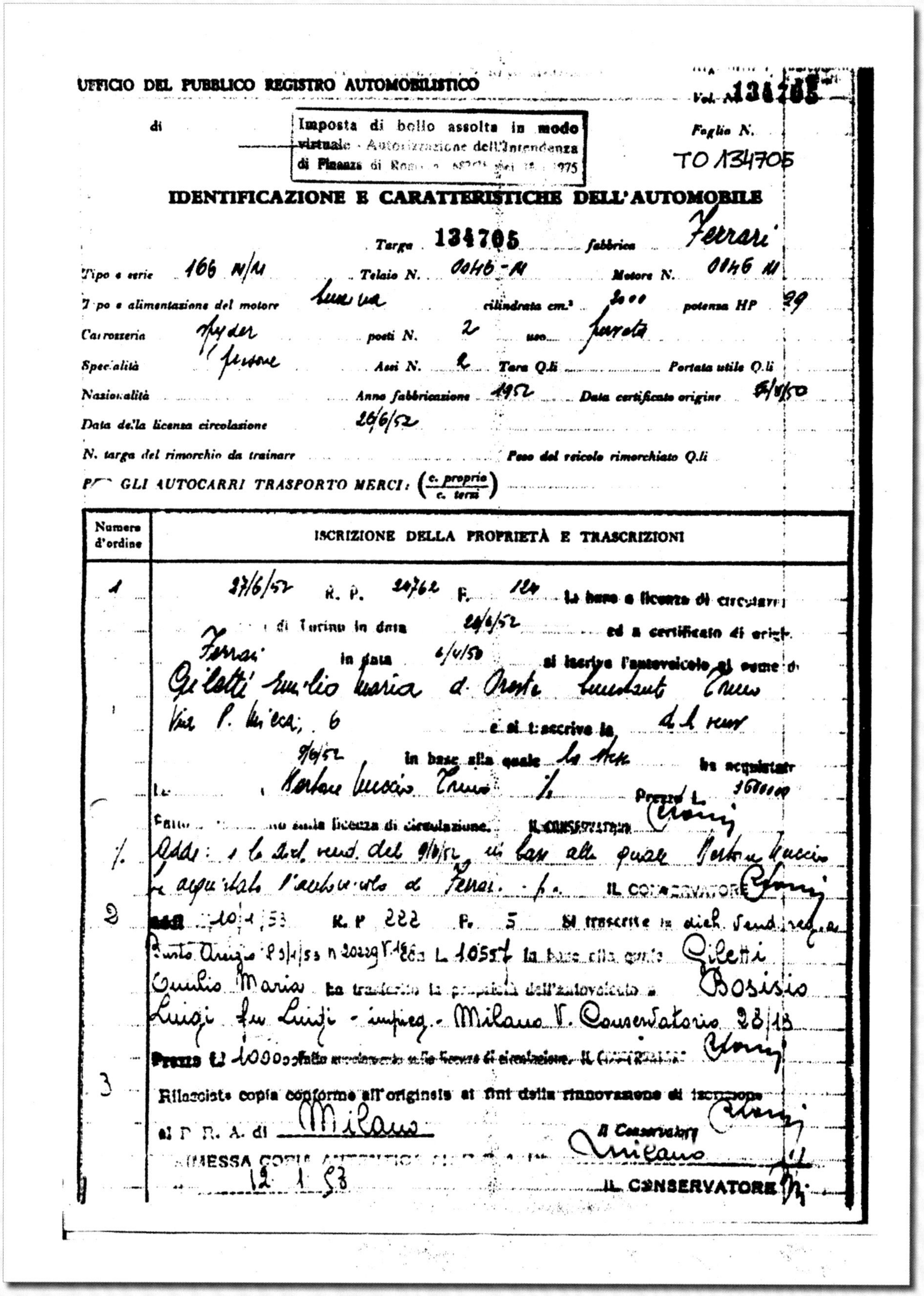

UFFICIO DEL PUBBLICO REGISTRO AUTOMOBILISTICO Vol. N. 134705

di

Imposta di bollo assolta in modo virtuale - Autorizzazione dell'Intendenza di Finanza di Roma ... 1975

Foglio N.

TO 134705

IDENTIFICAZIONE E CARATTERISTICHE DELL'AUTOMOBILE

Targa 134705 fabbrica Ferrari

Tipo e serie 166 MM Telaio N. 0046-M Motore N. 0046 M

Tipo e alimentazione del motore benzina cilindrata cm.³ 2000 potenza HP 29

Carrozzeria spyder posti N. 2 uso privato

Specialità trasporto persone Assi N. 2 Tara Q.li ... Portata utile Q.li ...

Nazionalità ... Anno fabbricazione 1952 Data certificato origine 6/4/50

Data della licenza circolazione 26/6/52

N. targa del rimorchio da trainare ... Peso del veicolo rimorchiato Q.li ...

PER GLI AUTOCARRI TRASPORTO MERCI: (c. proprie / c. terzi)

Numero d'ordine	ISCRIZIONE DELLA PROPRIETÀ E TRASCRIZIONI
1	27/6/52 R. P. 20762 F. 124 In base a licenza di circolare ... di Torino in data 26/6/52 ed a certificato di origine Ferrari in data 6/4/50 si iscrive l'autoveicolo al nome di Giletti Emilio Maria di Oreste ... Torino Via P. Micca, 6 e si trascrive la ... in data 9/6/52 in base alla quale lo stesso ha acquistato da Bertone ... Torino ... Prezzo L. 1600000 Fatto ... sulla licenza di circolazione. IL CONSERVATORE
1.	Add.: ... vend. del 9/6/52, in base alla quale Bertone ... ha acquistato l'autoveicolo da Ferrari ... IL CONSERVATORE
2	... 10-1-53 R. P. 222 F. 5 Si trascrive ... Santo Arcangelo 8/1/53 n 20229 ... L. 10550 in base alla quale Giletti Emilio Maria ha trasferito la proprietà dell'autoveicolo a Bosisio Luigi fu Luigi - impieg. - Milano V. Conservatorio 28/13 Prezzo L. 1000 ... Fatto annotamento sulla licenza di circolazione. IL CONSERVATORE
3	Rilasciata copia conforme all'originale ai fini della rinnovazione di iscrizione al P. R. A. di Milano Il Conservatore Milano ... RIMESSA COPIA AUTENTICA ... 12 1 53 IL CONSERVATORE

Appendix 3: Bertone Purchase and Giletti Registration Turin, page 2

Numero d'ordine

ISCRIZIONE DELLE IPOTECHE ED ANNOTAZIONI AD ESSE RELATIVE

30 Dic. 1953

di avviso

21.1.53 206 la rinnovazione

di Milano avvenuta in data 21.1.53 al n. 2065

Vol. 2064 Serie I

Il Conser[illegible]

Conservatore

SFORZA

Appendix 4: Bosisio Registration Milan

Reg. Serie 1ª (autoveicoli)
Vol. N. 2065
Foglio N. 206517

206517

AUTOMOBILE CLUB D'ITALIA

UFFICIO DEL PUBBLICO REGISTRO AUTOMOBILISTICO

di MILANO

IDENTIFICAZIONE E CARATTERISTICHE DELL'AUTOVEICOLO

Numero della licenza di circolazione 206517 *Fabbrica* Ferrari
tipo e serie 166/MM *telaio N.* 0046 M *motore N.* 0046 M
scoppio diesel elettrico } *alimentaz.* Benz. *cilindrata (cm³)* — *potenza (HP)* 89
carrozzeria Spider *posti N.* 2 *uso* priv.
velocità — *assi N.* — *tara q.li* — *portata utile q.li* —
nazionalità Ital. *se nuovo: anno fabbricaz.* — *data certificato origine* 6.4.50
data della licenza di circolazione 20.1.53
PER LE TRATTRICI: *peso del veicolo rimorchiato q.li* — PER LE MOTRICI: *N. targa del rimorchio da trainare* —
PER I RIMORCHI CON FRENO CONTINUO: *N. targa motrice autorizzata al traino* — PER GLI AUTOCARRI: *trasporti merci* c. proprio / c. terzi

Numero d'ordine	ISCRIZIONE DELLA PROPRIETÀ E TRASCRIZIONI
1	Addì 21.1.1953 Reg. Proc. N. 3970 Fasc. 20 tratto dalla copia autentica del foglio N. 134705 Serie 1ª Vol. / rilasciata in data 12.1.53 dal P.R.A. di Torino
	1) 2.6.52 N. 24762 .. 124 certif. origine e dichiarazione di vendita
	Giletti Emilio Maria di Oreste - Torino
	2) Addì 10.1.53 N. 888 F. 5 TRASCR. PRIV. SCRITT. AUT. 3.1.53 — DI VENDITA
	a Bosisio Luigi fu Luigi - Milano // Cons. - Fco
2	Sulla base della copia sopra trascritta - della nuova lic. circol. n. 206517 Mi - rilasciata il 20.1.53 — si rinnova la iscrizione dell'autoveicolo e la trascrizione della proprietà a nome di Bosisio Luigi fu Luigi - Milano - Via Conservatorio 28/13 - [illegible] IL CONSERVATORE [signature]
3	Addì 11.12.53 N. 86308 F. 432 si trascrive la dich. di vend. del 9.12.53 aut. a Broggi il 10.12.53 N. 18138 Vol. 162 atti Priv. L. 10.557 per il trasferimento di proprietà da: Luigi Bosisio a Martignoni Bruno fu Enrico

Appendix 5: Martignoni Registration Varese, page 1

Reg. Serie I^a (automobili)
Vol. N. 400
Foglio N. 39901

Imposta di bollo assolta in modo virtuale
Autorizzazione dell'Intendenza di Finanza
di Roma n. 9123... del 28 - 11 - 1973

39901

AUTOMOBILE CLUB D'ITALIA

UFFICIO DEL PUBBLICO REGISTRO AUTOMOBILISTICO di Varese

IDENTIFICAZIONE E CARATTERISTICHE DELL'AUTOVEICOLO

Numero della licenza di circolazione 39901 VA già N. 206517
Fabbrica Ferrari *Tipo e serie* 166 MM
Telaio N. 0046 *motore N.* 0046 M
alimentaz. benz. *cilindrata (cm³)* — *potenza (HP)* 29
carrozzeria Spyder *posti N.* 2
uso priv.
specialità — *assi N.*
tara q.li — *portata utile q.li*
nazionalità ital. *anno fabbricaz.* 1950 *data certificato origine* 6-4-50
data della licenza di circolazione 26-3-54
PER LE TRATTRICI: *peso del veicolo rimorchiato q.li*
PER LE MOTRICI: *N. targa del rimorchio da trainare*
PER ~~GLI AUTOCARRI~~: *trasporti merci* c. proprio / c. terzi

Numero d'ordine	ISCRIZIONE DELLA PROPRIETA' E TRASCRIZIONI
	Dalla copia aut. PRA di Milano fog. 206517
1	Roldi 21-1-53 RP. 3970 f. 20 Estratto dalla copia aut. fog. 134705 ril. 12-1-53 PRA di Torino
1	27-6-52 RP. 24762 f. 124 In base a certif. orig. e di chiar. vendita iscritta a Giletti Emilio Maria di Oreste Torino –
2	10-1-53 RP. 888 f. 5 Trasfer. a Bosisio Luigi fu Luigi, Milano
3	Sulla base della copia sopra trascritta e della nuova lic. di circol. n. 206517 MI – ril. 20-1-53 si rinnova la iscrizione a fav. di Bosisio Luigi fu Luigi, Milano
4	11-12-53 RP. 86308 f. 432 Si trascrive dich. vendita 9-12-53 reg. a Busto 10/12/53 n. 18638 Vol. 162 Atti Priv. L. 10.557.- per il trasf. propr. da Luigi Bosisio a Martignoni Bruno fu Enrico – Varese – Via Cavour 42. Vebom.

– 1 –

Appendix 5: Martignoni Registration Varese, page 2

Numero d'ordine	ISCRIZIONE DELLA PROPRIETA' E TRASCRIZIONI
1	Addì 7-4-54 R. P. 3085 F. 16 Sulla base della ... aut. sopra trascritta e nuova licenza di circolaz. n. 39901 rilas. 26/3 dalla Prefettura di qui, si rinnova l'iscriz. della proprietà ... a favore di Martignoni Bruno fu Enrico - Varese - Via Cavour 42 Valore dichiar. L. – Fatto annot. sul ... CONSER.
2	Addì 13·9·55 Rp. 9439 F 48 In base a scheda di denuncia del 13·9·55 rilasciata da Martignoni Bruno - Varese si annota la radiazione dell'autov. retro descritto perchè demolito. Il Conservatore

PUBBLICO REGISTRO AUTOMOB. A.C.I. VARESE -2-8-91

PUBBLICO

Appendix 6: Varese Registration Booklet

Source: Davide del Medico

Appendix 6: Varese Registration Booklet

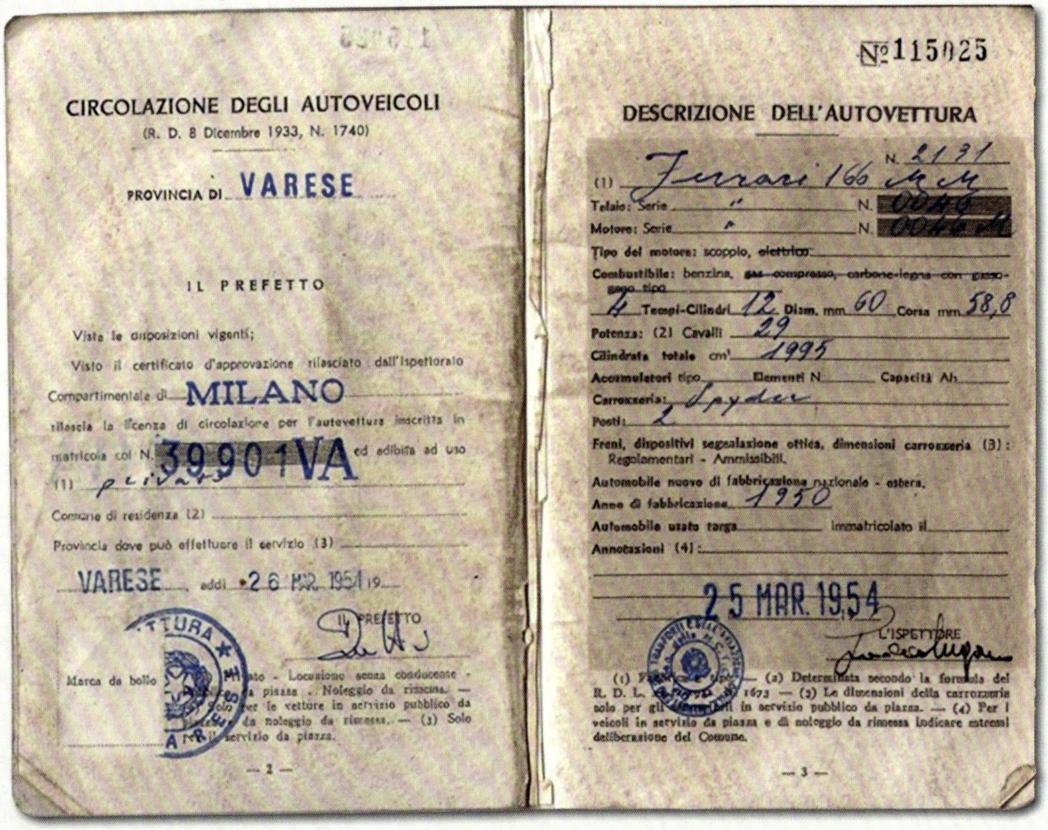

CIRCOLAZIONE DEGLI AUTOVEICOLI
(R. D. 8 Dicembre 1933, N. 1740)

PROVINCIA DI VARESE

IL PREFETTO

Viste le disposizioni vigenti;
Visto il certificato d'approvazione rilasciato dall'Ispettorato Compartimentale di MILANO rilascia la licenza di circolazione per l'autovettura inscritta in matricola col N. 39901VA ed adibita ad uso (1) privato
Comune di residenza (2)
Provincia dove può effettuare il servizio (3)
VARESE addì 26 MAR. 1954 19
IL PREFETTO

Marca da bollo

— 2 —

№ 115025

DESCRIZIONE DELL'AUTOVETTURA

(1) Ferrari 166 MM N. 2131
Telaio: Serie " N. 0046
Motore: Serie " N. 0046 M
Tipo del motore: scoppio, ~~elettrico~~
Combustibile: benzina, ~~gas compresso, carbone-legna con gassogeno tipo~~
4 Tempi-Cilindri 12 Diam. mm 60 Corsa mm 58,8
Potenza: (2) Cavalli 29
Cilindrata totale cm³ 1995
Accumulatori tipo Elementi N. Capacità Ah
Carrozzeria: Spyder
Posti: 2
Freni, dispositivi segnalazione ottica, dimensioni carrozzeria (3): Regolamentari - Ammissibili.
Automobile nuovo di fabbricazione nazionale - estera.
Anno di fabbricazione 1950
Automobile usato targa immatricolato il
Annotazioni (4):
25 MAR. 1954
L'ISPETTORE

(1) ... — (2) Determinata secondo la formula del R. D. L. ... 1673 — (3) Le dimensioni della carrozzeria solo per gli ... in servizio pubblico da piazza. — (4) Per i veicoli in servizio da piazza e di noleggio da rimessa indicare estremi deliberazione del Comune.

— 3 —

GENERALITÀ DEL PROPRIETARIO
Sua residenza e relative variazioni. Annotazioni degli eventuali passaggi di proprietà e relativi cambiamenti di targa, quando sia il caso.

Martignoni Bruno fu Enrico
via Cavour 42
Varese
IL PREFETTO
26 MAR. 1954

IL PREFETTO
il 19

IL PREFETTO
il 19

— 8 —

№ 115025

GENERALITÀ DEL PROPRIETARIO
Sua residenza e relative variazioni. Annotazioni degli eventuali passaggi di proprietà e relativi cambiamenti di targa, quando sia il caso.

IL PREFETTO
il 19

IL PREFETTO
il 19

IL PREFETTO
il 19

— 9 —

Appendix 7: Shipping Manifest to New York, page 1

TELEPHONES:
'TEHALL 3-7331 TO 7348
(18 LINES)

CABLE
"TRAMARSHIP"

INTRA-MAR TRANSPORT CORP.

CUSTOM HOUSE BROKERS

42 STONE STREET – NEW YORK 4, N. Y., U. S. A.

LUIGI CHINETTI
252 EAST 61ST STREET
NEW YORK, NEW YORK

REF. NO. 71894 — DATE SEP 19 56 — INV. NO. 32520

CARRIER: CHRISTOFORO COLOMBO

SHIPPER: M.G. CREPALDI AUTOMOBILE, MILAN, ITALY

FORWARDER: JMC GENOA P 12655

MARKS & NUMBERS: 1 MASERATI AUTOMOBILE UNBOXED WITH 5 WHEELS & TIRES -MOTOR 083- FRAME 083
1 FERRARI AUTOMOBILE UNBOXED WITH 5 WHEELS & TIRES MOTOR 0046 - FRAME 0046

It is agreed that if the property covered by this bill is entrusted or delivered to any express company, truckman, steamship, railroad or other carrier, whether named herein or not (which Intra - Mar Transport Corp. are hereby authorized to do, subject to all the usual conditions of transportation of such carriers or others) the shipper, consignee or owner of said property shall not demand or recover from Intra - Mar Transport Corp., any amount in excess of $50.00 per package (or the invoice value if less) except by special written agreement.
A bond has been given on your account guaranteeing the re-delivery to the Customs of unexamined goods not conforming to the various regulations. Goods should not be disposed of until packages retained for examination have been passed.
With the exception of payment of duties, in handling this shipment, we are acting as an independent contractor.

Shipper's C.O.D. $	
Foreign charges INLAND FREIGHT AND EXPENSES TO F.O.B. GENOA $	183.47
C.O.D. collection fee, cable advice & remittance $	
Ocean freight $	475.00
Transfer of documents, service and accounting $	
Duty as per attached $	204.40
This is a deposit only, subject to change on liquidation of the entry based on the Appraiser's report. If you have any other invoice or document showing a different value from that shown thereon, advise us immediately.	
CONSUMPTION Entry and formalities $	35.00
Entry and services $	
Entry bond and services $	2.50
Cartage and services $	12.50
COOPER'S EXPENSES	
Handling, reforwarding, B/L, etc. $	
Insurance & Services $	
Submission of invoice to appraiser $	
Storage and labor at F.T.Z./warehouse $	
Blanks, stamps, notary fees, petties $	
Special messengers $	20.00
Special service regarding delivery $	
Bond for consular invoice $	
Deposit for non production of consular invoice, refundable if produced within 6 months of entry date $	
	933.67

All Shipments handled subject to the terms and conditions set forth in our Acknowledgment of Shipping Instructions printed hereon.

Terms cash.—The above items represent principally cash paid or to be paid out for your account, and include in each case our profit or compensation for our services rendered. Prompt remittance is therefore respectfully requested. Unless the shipment is entered in Customs within 48 hours after arrival of the steamer, the shipment will be placed in general order warehouse. It is important, therefore, that prompt remittance is received for the amount of this bill so as to permit us to pay duty and make the entry.

Appendix 7: Shipping Manifest to New York, page 2

Statement of Importers addition or deduction from invoice to make market value — Sec. 487 Tariff Act. 1930

5334o-MP

Item	Quantity	Invoice Unit of Value	Addition to Invoice Unit of Value	Deduction From Invoice Unit of Value	Entered Unit of Value	Total Added	Amount Deduction

Importer enters to add as follows:

1 used Maserati motor car AG 1500, complete w/5 wheels, & tires @ HM. Lira 700000.-
1 used Ferrari motor car 166/M, complete w/5 wheels & tires " " 540000.-

Lira 1240000.-
plus 3% tax 37200.-
Lira 1277200.-
@ .00160017--
$2043.74
Instead of as Invoiced: 2000.00
Adds $ 43.74

Appendix 8: 1965 Bill of Sale for the Engine

CHATEL MORTGAGE AND CONDITIONAL BILL OF SALE

Birmingham Jefferson County State of Alabama Date: AUGUST 2, 1965

Know all men by these presents that in consideration of the sum of $ 50.00 and other value received, in hand paid by DARRELL O. WESTFAUL (BUYER) and the balance of $325.00 to be paid in regular installments within 12 months, or paid on demand by AUGUST 1, 1966, I do hereby sell all my rights, title, and interest in one automobile engine, to wit:

One (1) Ferrari Automobile engine - Type 166 Model 1954
Serial No: 166-10324

The above engine may be delivered on receipt of a total receipt of $187.50, and the balance of $187.50 to be paid after delivery. The Buyer DARRELL O. WESTFAUL agrees that if total payment is not made within 12 months that he will pay any reasonable attorney's fees and necessary court cost for collection of said debt.

I, Robert V. Rabun (SELLER), hereby further certify that the above named automobile engine is entirely free and clear of liens, and all encumbrances.

BUYER TAKES POSSESION OF ENGINE AFTER 50% OF TOTAL IS PAID (RVR)

SIGNED Robert V. Rabun SELLER
Sellers Address 7101 DIVISION AVE.
City of Birmingham, Alabama

SIGNED: Darrell O. Westfaul BUYER
Buyers Address 151 GLENWOOD STREET
City of Mobile, Alabama

NOTORIZED:

STATE OF ALABAMA,
Jefferson County

Sworn to and subscribed before me this the 2 day of Aug 1965

Hugh [illegible] Notary Public

My Commission Expires NOV 1966

Appendix 9: 1966 Bill of Sale for the Car

BILL OF SALE

Know all by these presents:

That in consideration of $1500.00 _______________________ in hand paid by DARRELL O. WESTFAUL, hereinafter referred to as buyer, the receipt of which is hereby acknowledged, the undersigned does sell and convey to the said buyer one Ferrari 166 Mille Miglia Motor No. T409D Serial No. 3447, and does hereby covenant and agree to warrant and defend the sale of said automobile unto the said buyer against any lawful claims and demands of all and every person or persons whatsoever.

Sold as is without warranty unless otherwise specified.

Witness my hand this 5TH day of July, 1966.

Witnesses: Conrad T. Harwell

Signed: Charles H. Deeman
Seller

P.O.B. 3102 East Side Station
Address

Tuscaloosa Ala.
City State

I hereby acknowledge receipt of above described automobile:

Darrell O. Westfaul
Buyer

151 Glenwood St.
Address

Mobile, Alabama
City State

Appendix 10: Last License Tag Receipts

Application and Transfer of 1966 Motor Vehicle License 63-27562

STATE OF ALABAMA, TUSCALOOSA COUNTY

To: Commissioner of License

I, the undersigned, hereby make application to you as provided by Sec. 706, Article 8, Title 51, Code of Alabama of 1940, Recompiled, to have the above license issued by you for:

Make	Year	Model	Body Type	Identification No.
Ferrari	54	166 Miglia	Cpe 8c	T 4094

which I (we) purchased, traded in, repossessed, or otherwise acquired,

from: Charles G. Deerman

Address: Box 3102 - East Side Sta. Tuscaloosa

To: Darrell O. Westfaul

Address: 151 - Glenwood St - Mobile

7-12-66 — **(Date Acquired)**

By application — **(Signature of applicant)**

This is to certify that, proper application having been made to me, I have this day transferred the above license.

Sworn to and subscribed before me this 12 **day of** July **, 1966.**

Commissioner of License

COPY FOR TRANSFEREE

By R. B. Berry, Lic **(Clerk)**

1966 Tag No. 63-27562

STATE OF ALABAMA
County of Tuscaloosa

1967 Tag No. 63-11025

1967 TAGS MUST BE DISPLAYED BEFORE NOVEMBER 16, 1966

THIS VEHICLE IS DESIGNED AND USED FOR PRIVATE PASSENGER PURPOSES

Valuation	State & County	Municipality	Sch. Dist. Tax	Total Tax	Lic. & Iss. Fee	Total	Mail Fee	Payable If Mailed
300	540	300	240	1080	375	1455		

Make	Year	Model	Body Type	Motor or Identification No.
Ferrari	54	166 Miglia	Coupe 8c	T 4092

Darrell O. Westfaul 3A
707½ - 10th St. Apt. 3
Tuscaloosa, Ala

Citation

Penalty

Total

ACQUIRED FROM NEW	USED	Pur. Price	Date Acquired	Date Ent. Alabama

The above vehicle was not operated on the highways of Alabama between ______ and ______.

I, The undersigned, hereby certify that the information hereon is true and correct. Sworn to and subscribed before me this 11 day of Oct 1966

Darrell O. Westfaul
Licensee or Applicant

R. B. BERRY
License Commissioner

By EB Betty
Clerk

MOTOR VEHICLE LICENSE AND TAX RECEIPT
No Title Certificate Required
1967 Tags Expire Sept. 30, 1967
(see reverse side)

Appendix 11: 1966 Letter from Ferrari S.p.A.

Ferrari
SOCIETÀ PER AZIONI ESERCIZIO FABBRICHE AUTOMOBILI E CORSE
CAPITALE SOCIALE UN MILIARDO

Ns.rif.SAT/GG/Bd n°1106   Modena,li 28 luglio 1966

Egregio Signor
DARRELL WESTFAUL
151 Glenwood St.
MOBILE, ALABAMA - U.S.A.-

OGGETTO:Vettura tipo 166 M.-

Riscontriamo la Sua del 13 luglio u.s. e nel foglio alle= gato alla presente Le trasmettiamo tutte le informazioni tecniche richieste.

Desideriamo precisarLe che il numero di matricola della vettura da Lei indicato (3447) è errato e che non abbiamo più disponibilità di particolari di ricambio.

Ci è gradito l'incontro per ben distintamente salutarLa.

Ferrari
S.p.A. SEFAC
Servizio Assistenza Clienti

All:1 -

SEDE SOCIALE - SERVIZI COMMERCIALE E ASSISTENZA CLIENTI: MODENA VIALE TRENTO TRIESTE 31 - TEL. 24.081 - 24.082 { CASELLA POSTALE: 232 MODENA
DIREZIONE - AMMINISTRAZIONE - GESTIONE SPORTIVA - OFFICINE: MARANELLO - TEL. 91.161 - 91.162 - 91.188 - TELEX - FERRARI - 51.013 { TELEGRAMMI: FERRARI - MODENA

Appendix 12: 1966 Typewritten Specifications from Ferrari, Page 1 of 6

VETTURA TIPO 166 M

Anno di fabbricazione 1950

Telaio 0046 M - Motore 0046 M

Carrozzeria Touring

Data di consegna 6 aprile 1950

1°- Proprietario - Sig.Bertone Nuccio - Torino

2°- Proprietario - Sig.Martignoni Bruno - Varese

N O T A

Trattandosi di vettura venduta indirettamente al Cliente,non ci è stato possibile seguire la sua attività sportiva.Benchè essa non figura tra quelle impiegate dalla Casa nelle gare in Messico e nel Nord America,non è da escludere che essa abbia partecipato ad alcu= ne di esse in quei Paesi.-

---

Caratteristiche principali della vettura alla consegna (1950)

MOTORE

Numero dei cilindri con V di 60° 12

Alesaggio e corsa mm. 60 x 58.8

Cilindrata totale cc. 2000 (1995)

Rapporto compressione 1: 8.5

Regime massimo 7000 giri

Potenza max CV. 136

Coppia max a 500 giri kgm. 13.6

Fasatura

Aspirazione - apre con 25° di anticipo
chiude con 53° di posticipo

Scarico - apre con 63° di anticipo
chiude con 16° di posticipo

Gioco alle valvole (con motore freddo)

aspirazione mm. 0.15
scarico mm. 0.20

Appendix 13: *Prancing Horse* Article

HORSING AROUND

(Editor's note: During the past few months two new members joined the FCA, each with an early Ferrari of some significance, and each of whom sent us the history of his car. We thought that their stories merited publication.)

A Short History of 0046M

by Darrell Westfaul

0046M *when owned by Chuck Saffel.*

Peter Briggs with 0060M.

I purchased 0046M on July 5, 1966 — it was my first car. Like so many other Ferraris of that time it had been given a Chevrolet engine. However, in August of 1965 I had already bought the original engine from another party with the idea of eventually joining the two again. Now, in 1988, the intention is still the same, but the project has been on hold for many years.

I have recently begun researching 0046M's history, and have pieced together the following from the literature and interviews available to me thus far:

According to a letter from Ferrari in 1966, 0046M was completed on April 6, 1950, for Nuccio Bertone. On April 23, 1950, Bertone raced a 166 MM, undoubtedly this one, in the Mille Miglia, finishing third in class and fourteenth overall. In 1952, Bertone entered the same car in the Mille Miglia again, but did not finish. A photograph of him in the 166 MM during the race shows two louvers in the front of the Ferrari's hood identical to those on 0046M; it is a hood design that may be unique to 0046M among the barchettas.

In 1952, 0046M was entered in the Mille Miglia by Emilio Giletti, who did not finish. Presumably driving the same 166 MM, Giletti had two firsts in class in June of that year at the Coppa della Toscana and the Giro dell'Umbria, finishing third overall in the latter. He finished first overall in September at the Trofeo Sardo, and was rewarded with mention in the 1952 Yearbook.

Bruno Martiguoni acquired the car after Giletti and had Zagato add a top which was smoothly contoured into the rear. This was probably finished in early 1953.

From 1953 through 1955 its whereabouts and activities are unknown.

In early 1956, John R. Sharpe of Buffalo purchased 0046M from Chinetti, I believe. Sharpe raced the car in the 1956 Watkins Glen Grand Prix with Fred Woodmansee driving. A spun wheel hub forced it to retire while leading the third place Testa Rossa from LeMans of that year.

Sharpe sold the car to someone who took it to Detroit; there it was sold to Chuck Saffel from Huntsville, Alabama, in 1957. Saffell owned and raced a number of Ferraris and his picture is in the 1957 Yearbook. After restoring the car, Saffel kept 0046M for several years, driving it everywhere, before selling it to Maurice Blevins of Birmingham.

Blevins is distinguished for having blown the engine at a drag strip on his first weekend of ownership. Then he installed the Chevrolet engine.

From Blevins, 0046M went to Charles Deerman of Tuscaloosa, Alabama, who did no real harm to it other than sandblasting to remove the paint, which also removed most of the filler. It was at that point, and in that condition, that I bought 0046M.

One very minor correction: The third place car at LeMans in 1956 was actually a 625 LM, serial number 0644. — Editor

Ferrari Type 195 Sport

It is believed only 36 type 195's were produced in total between 1950 and 1951. Only 10 of the short wheelbase Le Mans Touring Berlinettas in total were ever put on Ferrari chassis and one of these is chassis 0060M, the vehicle on display.

This car was first shown at the Geneva Auto Show in March 1950, when it was announced that it was intended for forthcoming long distance races, including the Targa Florio, the Mille Miglia and Le Mans.

On April 2, 1950 the car was entered in the Targa Florio Tour of Sicily and, driven by Albert Ascari, led for a third of the race before retiring. The car was next entered for the Mille Miglia on April 23, 1950, driven by Gianinni Marzotto. Despite the very bad weather conditions, the car won the event, averaging 76.6 mph (124 kph).

Le Mans was next and the car was entered, driven by Raymond Sommer and Dorino Serafini. Wearing race number 25, the car was fastest in the trials and assumed the lead on the first lap, having passed at least a dozen cars on that lap. The pace proved too much and at the sixth hour, the car was retired with a broken generator mounting bracket.

Further successes in Europe included Paolo Marzotto's wins in the Dolomite Cup and the Gallenga Cup, and Dorino Serafini's win in the Tour of Calabria.

On the last day of 1950 the car finished 1st in class and 5th overall in the Sebring Six Hour Race, becoming the first closed Ferrari to arrive in the USA. The car was driven by Luigi Chinetti the whole six hours of the race without a co-driver, after he had already driven the car from New York to Florida for the race.

In 1951, the car was purchased by Briggs Cunningham, who entered it in the Juan Peron Sports Car Grand Prix in Buenos Aires and was rewarded with a class win. The driver was Jim Kimberly.

May, 1951 saw the car finish 1st in class and 2nd overall in the Mecox Trophy Race on Long Island, driven by Phil Walters. At Watkins Glen the car again finished 1st in class and 2nd overall in the Seneca Cup, driven by John Fitch and still painted blue.

When Cunningham committed his resources to the cars bearing his name, the Ferrari was sold to the Yung brothers of Long Island. Luigi Chinetti then took the car in trade from them and it was sold to Henry N. Manney III of *Road & Track* magazine fame; he also raced it on the west coast of America.

Manney subsequently sold the Ferrari to John Fox who raced it at Torrey Pines. It then passed through a succession of owners, including a repurchase by Henry Manney. The car at this time was in need of work to restore it and this was commenced by John Cannon, a friend of Manney. Russell Martin purchased it from Cannon and painted the Ferrari red. Gerald Willburn of Hawthorne, California saw the car and it became his. He redid the front end after it had been damaged while parked.

Willburn in turn sold the Ferrari to Larry Knaack who, after ten years of ownership, sold it to John Hadjuk of Chicago.

The last known USA owner was Don Dethlefsen of Chicago who purchased the Ferrari twelve months later. The car spent over twelve years in his hands, mostly in storage, but a color booklet detailing the car's history was produced by Dethlefsen in 1977.

The car had remained in America for over 25 years when it was offered for sale in England, where it was purchased by the famous collector Peter Agg and finally sold by Coys of Kensington in 1980 for display at the York Motor Museum.

While 0060M has an interesting and significant history, it is our understanding that some of the early European history (Mille Miglia, Le Mans, etc.) cannot be absolutely attributed to 0060M. — Editor

Appendix 14: Michele Marchianò Letter from Zagato

ZAGATO CAR S.R.L.
stabilimento e uffici Amministrativi
20017 Terrazzano di Rho (Milano) Italia
Tel. (02) 9301635 Telex: 332528 ZAGATO I
Telefax (02) 9310285 ZAGATO I

car

Cod. Fiscale IVA 01257260164
Sede Soc. in Bergamo - Cap. Soc. L. 1.380.000.000
Tribunale di Bergamo - Reg. Soc. 16937 Vol. 15986
C.C.I.A.A. Bg. 187638 - C.C.I.A.A. Mi. 1058551

Mr. DARRELL WESTFAUL
P.O. BOX 2602
TUSCALOOSA, ALABAMA 35403
U.S.A.

DATA 15 Maggio 1989

Ns. Rif.

Vs. Rif.

Caro Amico

il mio ritardo nel rispondere alla tua lettera é imperdonabile ma volevo inviarTi la foto della tua bella Ferrari controfirmata da Anderloni e da Zagato. Purtroppo non sono riuscito ad incontrarmi con il primo e alla fine ho deciso di mandarti solo l'autentica del secondo.

Ti ringrazio per il materiale che mi hai mandato. Io, come vedi, faccio altrettanto allegandoti una rivista con un mio articolo e una tua foto.

Ora sto preparando un nuovo libro dedicato a tutte le vetture sportive di Zagato. Naturalmente inseriro' anche la tua Ferrari assieme ad altri 69 modelli da corsa.

Nei mei programmi c'é poi un catagolo generale Zagato per il quale avrei bisogno delle immagini della 0046 M dopo il restauro, possibilmente diapositive.

Per una serie di circostanze che non potevo prevedere il raduno dei settànt'anni Zagato è stato rinviato. Appena verrà presa una decisione Ti informero'.

Per quanto riguarda infine il figurino della 166 ti allego una fotocopia. Il proprietario chiede 3.000 dollari: mi sembra caro.Comunque fammi sapere.

Appena avro' altre notizie Ti scrivero'. Conto di incontrare nuovamente il Sig. Bosisio nei prossimi giorni e se avesse trovato altre foto della sua vettura provvedero' a fartele avere. La tua lettera mi ha fatto veramente piacere. Ho solo il cruccio di non riuscire mai ad essere tempestivo a causa dell'infinità di impegni di lavoro che mi assilla.

Un caro saluto a te a Mary Virginia Reese

Michele Marchianò

RECEIVED 7-20-89

Appendix 15: Emilio Giletti Fax, Page 1

JUL-28-'90 23:50 ID:GILETTI SPA PONZONE TEL NO:777104 #836 P01/06

Giletti

S.p.A. - Capitale Sociale L. 1.400.000.000 interamente versati
FONDATA NEL 1884

NS. FAX: (015) 777104

FAX NR. 374 ba DEL 27.07.90

PAGINA NR.
(totale pagine 6)

ATTENZIONE EGR. SIG. DARRELL WESTFAUL

RINGRAZIO PER SUO FAX DEL 28 FEBBRAIO 1990 E MI SCUSO PER IL RITARDO CON IL QUALE RISPONDO, COMUNQUE NON DIPENDENTE DALLA MIA VOLONTA'.

SONO PASSATI TANTI ANNI MA CERCHERO' TUTTAVIA DI RISPONDERE NEL MIGLIOR MODO POSSIBILE ALLE SUE DOMANDE.

1) OLTRE AD ESSERE UN PILOTA, PER LA MIA GRANDE PASSIONE PER LA MECCANICA VENIVO ANCHE IMPIEGATO COME COLLAUDATORE. EFFETTUAVO QUINDI TESTS E COLLAUDI SU MODIFICHE O NUOVE APPLICAZIONI CHE, SE I RISULTATI ERANO POSITIVI, VENIVANO POI UTILIZZATE PER LE NUOVE PRODUZIONI. NEL 1952 ABBIAMO EFFETTIVAMENTE COLLAUDATO CON ESITO POSITIVO UNA TESTA CILINDRI DIVERSA, CHE A PARTIRE DAL 1953 E' STATA UTILIZZATA ANCHE PER GLI ALTRI MOTORI.

2) LE NUOVE TESTE CILINDRI MI SONO STATE MONTATE DALLA FERRARI.

3) VEDI PUNTO 1

4) LE INSTALLAZIONI SONO STATE FATTE DALLA SCUDERIA GUASTALLA, DIETRO ALLA QUALE C'ERA COMUNQUE LA FERRARI.

5)- 6) TUTTE LE MODIFICHE VENIVANO FATTE DALLA FERRARI PER ESPERIENZA. CON QUESTA MACCHINA NEL 1952 HO VINTO IL CAMPIONATO ITALIANO CATEGORIA SPORT.

7) - 8) SONO PASSATI ORMAI TROPPI ANNI E NON SONO IN GRADO DI DARLE QUESTE NOTIZIE.

9) HO INIZIATO A CORRERE A 18 ANNI E HO FATTO LE PRIME CORSE IN SALITA CON LA LANCIA TIPO ARDEA; LA MIA PRIMA CORSA E' STATA LA BIELLA-OROPA. HO POI PARTECIPATO AI RALLY CON LA FIAT MORETTI

SE I DOCUMENTI RISULTANO ILLEGGIBILI, AVVISATECI

segue

GILETTI

SEDE LEGALE DIREZIONE E STABILIMENTO: VIA PROVINCIALE 51 - 13064 PONZONE BIELLESE (VERCELLI)
TEL. (015) 77.70.41-42 – TELEGR. GILETTI PONZONE-TRIVERO – TELEX 222427 GILOR I
C.C.I.A. VERCELLI N. 27185 – ISCR. TRIB. BIELLA REG. SOC. N. 4780 – STAZIONE FF. SS. BORGOSESIA
COD. FISC. E PART. IVA N. 0018133 002 2 – C/C POSTALE N. 11008137 – POSIZ. EXPORT M. 891.798

Appendix 15: Emilio Giletti Fax, Page 2

Giletti

S.p.A. - Capitale Sociale L 1.400.000.000 interamente versati
FONDATA NEL 1884

NS. FAX: (015) 777104

FAX NR. 374 ba DEL 27.7.90 PAGINA NR. 2
(continuazione)

750.
HO VINTO LA MIA PRIMA GARA, LA VARESE-CAMPO DEI FIORI, IL 26 GIUGNO 1951, CON LA N.D. 750, CON LA QUALE SEMPRE NEL 1951 HO VINTO DIVERSE CORSE IN SALITA.
SUCCESSIVAMENTE HO PARTECIPATO AD ALTRE CORSE PRIMA CON LA LANCIA APRILIA E POI CON LA LANCIA AURELIA B21, CON LA QUALE SONO ARRIVATO TERZO ASSOLUTO AL 12° GIRO DI SICILIA, IL 9 MARZO 1952.
COME LEI GIA' SA, SONO POI PASSATO ALLA FERRARI E NEL 1953 ALLA MASERATI.

10) LAVORAVO NELL'AZIENDA TESSILE DI PROPRIETA' DI MIO PADRE.

11) HO ACQUISTATO LA FERRARI 2000 DA NUCCIO BERTONE, DEL QUALE SONO MOLTO AMICO.

12) HO PARTECIPATO AD UNA GARA A MONZA NEL 1952. SONO STATO COSTRETTO AL RITIRO PER NOIE AL MOTORE AL 12°GIRO, COMUNQUE DOPO AVER OTTENUTO IL MIGLIOR TEMPO ASSOLUTO SUL GIRO E MENTRE MI TROVAVO IN TESTA ALLA GARA.
SEMPRE NEL 1952 HO VINTO CON IL MIGLIOR TEMPO ASSOLUTO E DI CATEGORIA IL TROFEO SARDO; IL MIO RECORD DI VELOCITA' ESISTE ANCORA.

13) 31 AGOSTO 1952: ENNA (GIRO DI SICILIA) TERZO ASSOLUTO
12 OTTOBRE 1952; CC BOLOGNA RATICOSA QUINTO ASSOLUTO E TERZO DI CATEGORIA

14) PERCHE' SONO USCITO DI STRADA A 40 KM. DALLA PARTENZA

15) LORO PIANA ERA IL MIO MECCANICO.

segue

SE I DOCUMENTI RISULTANO ILLEGGIBILI, AVVISATECI

GILETTI
DIVISIONE TESSILE

SEDE LEGALE DIREZIONE E STABILIMENTO: VIA PROVINCIALE 51 - 13064 PONZONE BIELLESE (VERCELLI)
TEL (015) 77.70.41-42 – TELEGR. GILETTI PONZONE-TRIVERO – TELEX 222427 GILOR I
C.C.I.A. VERCELLI N. 27185 – ISCR. TRIB. BIELLA REG. SOC. N. 4760 – STAZIONE FF. SS. BORGOSESIA
COD. FISC. E PART. IVA N. 0018133 002 2 – C/C POSTALE N. 11008137 – POSIZ. EXPORT M. 891.798

DIVISIONE MECCANICA

Appendix 15: Emilio Giletti Fax, Page 3

Giletti

S.p.A. - Capitale Sociale L. 1.400.000.000 interamente versati
FONDATA NEL 1884

NS. FAX: (015) 777104

FAX NR. 374 ba DEL 27.07.90 PAGINA NR. 3
(continuazione)

16) TARGA FLORIO 1952: NON MI RICORDO BENE MA PROBABILMENTE MI SONO RITIRATO PER GUASTO AL MOTORE.

17) LE FARO' AVERE A SETTEMBRE COPIA DELLE FOTOGRAFIE DELLA MACCHINA.

18) NO. SONO PASSATI TROPPI ANNI. PROVI PRESSO GLI ARCHIVI DELLA SCUDERIA GUASTALLA, CHE UNA VOLTA AVEVA LA SEDE A MILANO, PRESSO IL RAPPRESENTANTE DELLA FERRARI CHE SI CHIAMAVA CORNACCHIA.

19) PER IL CAMPIONATO DEL 1953 NON HO AVUTO IL TROFEO. HO VINTO INVECE IL CAMPIONATO ITALIANO CATEGORIA SPORT NEL 1952, SEMPRE ALLA GUIDA DELLA FERRARI 2000. HO RICEVUTO UN TROFEO MA NON HO LA FOTOGRAFIA.

20) HO VENDUTO LA FERRARI 2000 AL SIG. BOSISIO PERCHE' MI AVEVA FATTO UN'OFFERTA MOLTO ALLETTANTE.

21) NO

22) NO

23) NO

24) HO CONOSCIUTO BRUNO MARTIGNOLI MOLTI ANNI FA E NON RICORDO DI PRECISO COSA FACESSE. ERA COMUNQUE SICURAMENTE UN PILOTA GENTLEMAN.

25) LE ALLEGO UN PICCOLO ELENCO DELLE CORSE ALLE QUALI HO PARTECIPATO.

RESTO A SUO DISPOSIZIONE PER OGNI ALTRA INFORMAZIONE DOVESSE OCCORRERLE. CORDIALI SALUTI. EMILIO GILETTI

SE I DOCUMENTI RISULTANO ILLEGGIBILI, AVVISATECI

GILETTI
DIVISIONE TESSILE

SEDE LEGALE DIREZIONE E STABILIMENTO: VIA PROVINCIALE, 51 - 13064 PONZONE BIELLESE (VERCELLI)
TEL. (015) 77.70.41-42 – TELEGR. GILETTI PONZONE-TRIVERO – TELEX 222427 GILOR I
C.C.I.A. VERCELLI N. 27165 – ISCR. TRIB. BIELLA REG. SOC. N. 4760 – STAZIONE FF. SS. BORGOSESIA
COD. FISC. E PART. IVA N. 00181330022 – C/C POSTALE N. 11008137 – POSIZ. EXPORT M. 891.798

DIVISIONE MECCANICA

Appendix 16: Jess Pourret Letter

JULY 1ST, 1990

MR. DARRELL WESTFAUL
P.O. BOX 032052
TUSCALOOSA, AL. 35405

DEAR MR. WESTFAUL,

I'M SORRY IT HAS TAKEN SO LONG TO ANSWER YOUR LETTER ABOUT 0046M. FIRST OF ALL IT CAME BY BOAT SO ONLY GOT TO THAILAND IN EARLY APRIL AND SECOND, MY FILES BEING IN FRANCE I HAD TO WAIT FOR MY VACATIONS HERE TO DO ANYTHING ABOUT IT! ANYWAY AT LONG LAST HERE'S WHAT COMES OUT OF IT. FIRST OF ALL LET ME THANK YOU FOR THE 166 PICS YOU SENT, QUITE INTERESTING AND HELPFUL. I WISH I WERE A BIT LESS LAZY AND DO A BIT OF WRITING ON THE 166'S, YOUR PICTURES REMIND ME OF THAT.
ANYWAY LET'S TALK ABOUT YOURS.
BAD (?) NEWS FIRST, 0046M DOESN'T LOOK LIKE IT DID THE '52 MM. E. GILETTI WAS RUMORED TO HAVE DRIVEN 0046M IN THE RACE BUT IT DOES LOOK LIKE HE DROVE 0044M. I HAVE PICTURES OF IT THERE AND IT'S <u>NOT</u> 0046M.

• • •

UNFORTUNATELY I DO NOT KNOW OF A GEARBOX AND THIS'S A BIG PROBLEM. A 212 BOX WOULD DO PERFECTLY BECAUSE IT'S THE SAME ONE SAVE FOR THE GEAR LEVER BUT EVEN A 212 BOX IS HARD TO FIND. YEARS AGO WHEN HARRAH WAS STILL ALIVE I SAW IN THE "RESERVE", A 166 MM MOTOR <u>WITH</u> <u>MAGS</u> STAMPED <u>0043S</u>! (DOESN'T SOUND RIGHT BUT IT WAS SO!) WHICH WAS MATED TO AN EARLY 166 GEARBOX (49/EARLY 50) – WONDER WHAT HAPPENED TO THIS. RONALD PINTO OF CALIFORNIA HAD THE CAR (0435) IN THE '80'S. ANYWAY IF I HEAR OF A G.BOX I'LL LET YOU KNOW OF COURSE.
WITH MY EXCUSES FOR BEING SO LATE IN ANSWERING AND WITH MY BEST REGARDS,

JESS G. [illegible]

Appendix 17: Chuck Saffell Letter

AMERICAN FEDERATION OF AVICULTURE INC.

To: Darrell Westfaul   Date; March 7, 1989

From: Chuck Saffell

Dear Darrell:

First, please forgive me for being so tardy in getting back to you, I have been ill and in and out of the hospital several times. I am enclosing the wheel, some photos of my cars and some other Ferrari, Maserati and Lamborghini, several of these are one off cars. The shots taken in front of my thop in the old days has several beauties especially the old Gran Prix which I loved to drive through town for gas, just to shake up the native's. Also have enclosed some SCCA ribbons, won with Ferrari and Maserati autos. I have also found several slides which include your car, I am having a print made and then will send the slides to you. Will also send some trophies won with Ferrari and a Ferrari helmet damaged in a flip (saved my head). (possibly)?

As soon as possible I will get down to storage shed for windscreen.

Best wishes to all,

C W Saffell

C. W. Saffell

Appendix 18: Invoice for 2007 Sale of 0046M

SHIPPING INVOICE

Shipped by **Darrell Westfaul**
906 31st Ave.
Tuscaloosa, Alabama

Shipped: **Ferrari 166 MM # 0046M**
And the following parts:

In the car:

1 Box--Windows
1 Box--Rear window and windows trim
1 Box--Instruments, switches, mirrors
1 Box--Carburetors, small trim
1 Box--Differential Housing
1 Box--Old carpets
1 Box--NOS universal joints (2)
1 Box--Cover plates
1 Box--Tail lights, odds and ends
1 Box--Fuel filler pipe and cap
1 Box--Starter, generator, and regulator
1 Can--Screws, nuts, and bolts
1 Belly pan
1 Transmission cover
1 Axle tube
1 Spare tire/wheel
Wood spacers and metal strap for fuel tank cover

Not in the car:

2 Wood crates--Engine parts
1 Wood crate--Engine block, sump, exhaust headers
1 Wood crate--Transmission
1 Box--Crankshaft
1 Box--Fuel tank

Consigned to Exotic Car Transport for shipping. Accepted by:
Loaded Car w/16 boxes
Contents Unknown

Driver
Date: June 11, 2007

CHRONOLOGY OF 0046M

1950

Date	No.	Event	Result	Driver
Mar 16	Motor 0046M assembly completed at Ferrari in Maranello.			
Mar 18	Motor dynamometer tested, results redacted from Assembly Sheets.			
Mar 29	Transmission No. 8 assembly completed.			
April 6	Certificate of Origin issued.			
April ?	Chassis 0046M at Carrozzeria Touring in Milan, body No. 3447 constructed on it, paint, trim.			
April 21	At Ferrari – drum brakes finished/adjusted. This date also appears to apply to the completion of the differential No. 0018 but that was needed earlier to have a rolling chassis for the bodywork.			
April	Loaned to Giuseppe "Nuccio" Bertone of Turin, Plexiglass wind wings added to outsides of windscreens.			
April 23	#652	XVII Mille Miglia	14th OA 3rd S2.0	Bertone/Luigi Casalegno
June 4	#1211	II Coppa della Toscana	5th OA 2nd S2.0	Bertone/Luigi Casalegno
July 4	At Ferrari – differential ratio changed to 9x42, matching transmission output.			
July 16	#147	IV Coppa d'Oro delle Dolomiti	11th OA 5th S+1.1	Bertone/Luigi Casalegno?
July 30		Susa-Moncenisio hillclimb	3rd S+1.1	Bertone
Aug 6	#52	Aosta-Gran San Bernardo hillclimb	2nd S+1.1 (License plate PROVA TO 99)	Bertone

1951

Date	No.	Event	Result	Driver
Feb 10	Steering box cover stamped 10-2-51.			
Mar 23	At Ferrari, motor converted to three Weber 36 DCF3 carburetors and differential returned to original 8x40 ratio, two slits cut in the front of the hood.			
April 1	#405	XI Giro di Sicilia	32nd OA 7th S2.0 (One center driving light)	Bertone/Mirko Landini
April 28-29	#346	XVIII Mille Miglia	DNF (At Fano, stone broke windshield, two driving lights)	Bertone/Mirko Landini
May	Returned to Ferrari.			
Aug 22	Steering box cover stamped REV 22-8-51 (revisione/inspection)			

1952

Date	No.	Event	Result	Driver
April	Loaned to Emilio Maria Giletti of Turin by Bertone.			
April 19	225-style 12 port heads with roller tappets and crankshaft ground to first undersize sent from Ferrari to be installed by Scuderia Guastalla of Milan.			
June 1	#1240	IV Coppa della Toscana	11th OA 1st S2.0 (Temporary license M 23742)	Giletti/Walter Loro-Piana
June 9	Sold to Bertone by Ferrari			
June 12	#304	IV Giro dell'Umbria	3rd OA 1st S2.0	Giletti/Walter Loro-Piana
June 26	Resale to Emilio Maria Giletti, 3,600,000 lire, with Scuderia Guastalla.			
June 27	Sale recorded in Turin, issued license plate TO 134705.			
June 29	#70	XXXVI Targa Florio	DNF (Lap 7 of 8, motor)	Giletti
Aug 3	#702	IV Giro della Calabria	DNS	Giletti
Aug 10	#15	VIII Circuito Automobilistico di Senigallia	3rd OA 3rd S2.0 (Lap Record)	Giletti
Aug 15	#18	12 Ore di Pescara	6th OA 4th S+2.0 (One center driving light)	Giletti/Sergio Sighinolfi

Sept 14	#10	II Trofeo Sardo	1st OA 1st S1.1-2.0	Giletti/Walter Loro-Piana
Sept 28	#73	VI Gran Premio di Bari	11th OA 5th S2.0 (Flat tire)	Giletti
Oct 12	#320	Bologna - Raticosa hillclimb	5th OA 3rd S2.0	Giletti
Oct 26	#200	II Coppa Sant Ambroeus Monza	DNF (Lap 12 of 15, motor, set fastest lap time)	Giletti
Nov	Champion of Italy in Sports 2.0 class, 3rd in Assoluto Sport (all classes).			

1953

Jan 3	Sold to Luigi Attilio Bosisio of Milan, with Scuderia Sant Ambroeus.			
Jan 10	Sale recorded in Milan			
Jan 20	Registered and issued license plate MI 206517.			
Jan-Mar	Converted to a berlinetta by Carrozzeria Zagato, with a disabled second wiper on the driver's side, windshield washers, knee brace installed, seats recovered and headliner installed using light colored cloth, probably tan.			
Mar 29		Corsa in Salita sulle Torricelle hillclimb	7th OA 4th S+1.1	Bosisio
Apr 25-26	#520	XX Mille Miglia	DNF (Before Ravenna, axle broke)	Bosisio/Pier Luigi Ercole
May 16		Scala di Giocca-Osilo hillclimb	2nd 1.1-2.0	Bosisio
May 24	#20	III Trofeo Sardo	11th OA 1st S2.0	Bosisio/Pier Luigi Ercole
May 31	#917	V Coppa della Toscana	28th OA 3rd S2.0	Bosisio/Pier Luigi Ercole
June 28	#92	Chiusaforte-Sella Nevea hillclimb	20th OA 6th S+1.1	Bosisio
Dec 9	Sold to Bruno Martignoni, Varese, possibly via Piero Carini.			
Dec 10	Sale recorded at Busto Arsizio near Milan.			
Dec 11	Record transferred to Varese, no license plate issued.			
Dec ?	Shipped to Morocco.			
Dec 20	#22	12 Heures de Casablanca	7th OA 3rd S.2.0 (Two driving lights)	Martignoni/Piero Carini

1954

Mar 7	#44	Criterium Automobile du Sénégal	5th OA 5th S+1.5	Martignoni
Mar 25	Returned to Italy, inspected at Varese.			
Mar 26	Registered and issued license plate VA 39901.			
April 7	Inspection and registration recorded at Varese.			
'54 or '55	Repainted dark metallic gray, aluminum trim added on sides and hood, same seats, no headliner.			

1955

Sept	Sold or consigned to M. G. Crepaldi Automobili S.a.s at Milan, then sold/shipped to Luigi Chinetti, Chinetti Motors, New York, NY.
Sept 9	Arrived USA for New York customs, declared value 540,000 lire.
Sept 13	Martignoni declared 0046M "demolito".
Very late	Sold by Chinetti to David Francis Leopold of Buffalo, NY, $2-2,500.
Unknown	Seats recovered and headliner installed using red naugahyde.

1956

March ? Sold to John Reynolds Sharpe, Buffalo, NY, $2,500, has NY license plate 83-48BL.

Sept 15 #87 Watkins Glen Grand Prix — DNF (Windshield struck by track asphalt) — Fred Woodmansee

Sept 29 #65 Harewood Acres, Ontario, Canada, Indian Summer Trophy — DNF (Right rear axle broke) — Fred Woodmansee

Oct ? Sold back to Leopold with broken axle.

Sold to Delavan Smith Lee of World Sports Car Center, Detroit, MI, with broken axle.

Axle replaced by Richard Londergan, Dick's Auto Repair, Detroit, MI.

1957

March ? Sold to Charles "Chuck" and Audrey Saffell of Huntsville, AL, $3000 less trade of Triumph TR2, has MI license plate EG 5037.

Spring Right rear axle tube replaced.

Summer-Fall Motor rebuilt with new rings, car repainted in red, engine bay remains dark gray, disabled wiper removed and filled in, driving lights and brackets added, prancing horse decals.

1958

Mar 22 Driven to Sebring 12 Hours, has AL license plate 47-27921.

June-July Sold to Maurice Kevion Bivens of Birmingham, AL, has a prosthetic left leg below knee.

1959

Unknown Motor rebuilt, probably for new pistons.

1960

Late Ferrari motor dropped valve, replaced with 283 Chevrolet motor, transmission and differential. Car repainted red again, including engine bay, windshield washer nozzles removed and filled in, new black carpet on floor.

1961

Nov Motor 0046M sold to Robert Rabun and Michael Carpenter of Birmingham, AL. Repairs abandoned, Rabun takes motor, Carpenter takes transmission. Rabun stores motor parts under his bed, Carpenter installs transmission in Lotus 11, which is sold and disappears.

1962

June Ferrari sold to Charles Deerman of Tuscaloosa, AL, for $1,500, has AL license plate 1-93934.

1965

Aug 2 Motor 0046M sold to Darrell Westfaul of Mobile, AL, by Rabun, $375.

1966

June ? Ferrari sandblasted for painting, removing much body filler, driving lights removed, has AL license plate 63-27562.

July 5 Ferrari sold to Darrell Westfaul of Mobile, AL, $1,500.

July-Aug Body filler replaced by Westfaul and car painted red, prancing horse decals added.

Sept Returned to Tuscaloosa, AL, 1966 University of Alabama parking decal affixed to windshield, license plate 63-11005.

1967

May Returned to Mobile, AL, partially dismantled and stored.

1988

Oct Returned to Tuscaloosa, AL, stored.

1990

Oct 18 Replacement transmission No. 07 purchased from Tohru Horinouchi of Tokyo, Japan, $5,000.

2007

Apr 11 Transferred to the Darrell Oliver Westfaul Charitable Remainder Unitrust, trustee University of Alabama, for annual disbursements upon sale, first to Westfaul, then to scholarships.

May 21 Sold to Scott Gauthier of Scottsdale, AZ, $1,375,000.

June Ferrari displayed at Scottsdale Automotive Museum, still partially dismantled.

2008

June Motor 0046M sent to Epifani Restorations, Berkeley, CA, for repairs.

2015

Ferrari displayed at Gauthier's G Collection venue, still partially dismantled.

2016

Jul 13 Epifani Restorations finished motor repairs, $109,000, parts shipped back not assembled.

Aug Consigned to Gooding & Company Pebble Beach Auction, car reassembled, motor assembled and installed with transmission but not running, seats recovered in black leather, est. value $6-8,000,000.

Aug 20 Sold as Lot 68 to Andreas Pohl of Marburg, Germany, for $5,445,000, including 10% premium.

Relocated to underground garage at the Pohl home, Marburg, Germany.

INDEX

CARS/VEHICLES

David Tomaro

EVENTS & RACES

MANUFACTURERS

MEDIA

ORGANIZATIONS/BUSINESSES

David Tomaro

David Tomaro

PEOPLE

David Tomaro

PLACES

Design:	Jodi Ellis Graphics
Printer:	Interpress Ltd., Hungary
Page Size:	230 mm x 280 mm
Text paper:	150 gsm Magno gloss
End paper:	170 gsm Surbalin Glatt Karminrot 6161
Dust jacket:	150 gsm Glossy Artpaper
Casing:	Foil stamping on front and spine, on Plomo Geltex, over 3 mm board
Main Body Text:	10 pt. Articulat CF Regular
Captions:	9 pt. Articulat CF Medium
Chapter Heads:	24 pt. Articulat CF Regular
Sub Heads:	9 pt. Articulat CF Bold